Searchial Marketing:
How Social Media Drives Search Optimization in Web 3.0

Dr. Alan Glazier

AuthorHouse™
1663 Liberty Drive
Bloomington, IN 47403
www.authorhouse.com
Phone: 1-800-839-8640

© 2011 Dr. Alan Glazier. All rights reserved.

No part of this book may be reproduced, stored in a retrieval system, or transmitted by any means without the written permission of the author.

First published by AuthorHouse 2/14/2011

ISBN: 978-1-4567-3892-1 (sc)
ISBN: 978-1-4567-3891-4 (e)
ISBN: 978-1-4567-3893-8 (dj)

Library of Congress Control Number: 2011901581

Printed in the United States of America

Any people depicted in stock imagery provided by Thinkstock are models, and such images are being used for illustrative purposes only.
Certain stock imagery © Thinkstock.

This book is printed on acid-free paper.

Because of the dynamic nature of the Internet, any web addresses or links contained in this book may have changed since publication and may no longer be valid. The views expressed in this work are solely those of the author and do not necessarily reflect the views of the publisher, and the publisher hereby disclaims any responsibility for them.

DEDICATION

For my wife, Laura, and sons Jacob and Ari,

for exercising extreme patience and understanding while I pursue my

obsessive interests and ventures.

DISCLOSURE

Dr. Glazier is founder and owner of Schedgehog.com, the first mobile application that helps doctors and clinics fill missed and cancelled appointments, helping increase their revenue. Dr. Glazier is a professional blogger for SightNation.com and a consultant for CooperVision, both of whom are used as examples of social media sites in his book. Dr. Glazier has listed several links to websites in which he participates as an affiliate. He works as a consultant for thereadedge.com and demandforce.com. Other than that, Dr. Glazier has no commercial interest in any of the websites or products covered in this book.

Information specifically geared toward healthcare organizations, clinics, hospitals, and pharmaceutical companies can be found in the appendices.

CONTENTS

INTRODUCTION		xi
FOREWORD		xv
CHAPTER 1:	THE SOCIAL WEB; STARTING YOUR CONVERSATION	1
CHAPTER 2:	THE "SOCIAL" SIDE OF SEARCHIAL	20
CHAPTER 3:	SEARCH ENGINE ELEVATION STRATEGIES	25
CHAPTER 4:	OFF-PAGE OPTIMIZATION	60
CHAPTER 5:	LINKING	73
CHAPTER 6:	BLOGGING AND CONTENT CREATION	83
CHAPTER 7:	SOCIAL MEDIA SUITES	100
CHAPTER 8:	SOCIAL BROADCASTING (LIFESTREAMING)	131
CHAPTER 9:	THE MOBILE WEB AND LOCATION-BASED SOCIAL NETWORKS (AKA PLACESTREAMING)	139
CHAPTER 10:	SOCIAL REVIEW SITES, SOCIAL INFLUENCE, LOCAL SEARCH, AND CUSTOMER SERVICE TOOLS	146
CHAPTER 11:	ANALYTICS—BENCHMARKING YOUR SUCCESS AND MONITORING COMPETITORS' EFFORTS	157
CHAPTER 12:	SEARCHIAL MARKETING AND EMERGING NEW MEDIA TOOLS	170
AFTERWORD		175
ABOUT THE AUTHOR		177
SEARCHIAL FOR MEDICAL ORGANIZATIONS		179
APPENDIX A:	THE FDA AND MEDICAL INFORMATION AVAILABLE ON THE INTERNET	181
APPENDIX B:	SOCIAL MARKETING FOR MEDICAL PRACTITIONERS AND PROFESSIONALS	183
APPENDIX C:	SOCIAL MARKETING FOR CLINICS AND HOSPITALS	193
APPENDIX D:	SEARCHIAL MEDIA FOR BIOTECH AND PHARMACEUTICAL CONCERNS	203
BIBLIOGRAPHY		213

INTRODUCTION

Searchial Marketing

At a trade conference, when I heard Dr. Alan Glazier telling an audience how he had built his business by leveraging social media, my ears perked up. Here was a small-business entrepreneur with limited marketing means, yet you could find his site listings on top of the all-powerful Google search page, for everyone searching eye- and vision-care keywords and phrases in his geographic region. He was also the founder of his own community—#badvisiondecision—on Twitter. Who could not be interested by that?

What I learned from hearing Dr. Glazier speak was transformative to me, not because his content was particularly new to me—I had already covered a similar topic in the updated version of my book *Emotional Branding*—but the ability of this small-business owner to find time to connect directly with his patients and clients, and engage them without compromising his personal or professional life, was a message everyone should listen to. Here is a doctor who made the decision to reach out to people and invite them to learn about important issues regarding their eyes and vision that might be crucial to their lives. He was proof that we are not living in a fishbowl anymore, and that the right emotional content finds its audience wherever people are. He was passionate, engaged, involved—he was the trusted factor behind his brand, and that was the most powerful message of all.

So I Listened

In front of leaders from the powerful vision and optical industry—from manufacturers, such as Luxottica, to national chains, such as Walmart and Walgreens—Dr. Glazier captured the attention of everyone. His relentless message was that there is another way to market, and brands should drop the traditional marketing methods to reach out to consumers in a more personal way. For me, it was even more inspiring when I realized that Dr. Glazier's greater point was that consumers are waiting for *you*, the brands, to talk to *them,* and that brands are not generally jumping at the opportunity provided by new media to be part of that conversation.

Everyone listened to this small businessman intently. With his own experience on hand, and results proudly presented on the large screen via his PowerPoint presentation, I imagined Dr. Glazier as David addressing the Goliaths in the room. In a social media revolution, where a brand's reputation is often made online, powerful brands—not unlike Goliath—could be destroyed by more nimble and imaginative new companies if they did not see in their future the new power bestowed on consumers by social engagement technologies. Dr. Glazier was showing how to leverage social media to acquire customers and keep customers informed and increase loyalty within new online communities seeking trusted answers—communities any large brands would envy.

Everybody Should Listen Too

His story, and the story he is now laying out in his book *Searchial Marketing,* shows how the world of branding has changed, and how the emotional connection people have with brands occurs, to a great extent, online via social media. This connection is crucial for corporations to understand. Imagine a few years back this young doctor with a few medical offices, and a practice he needs to build, stressing about how he will grow his business and support his family. What do you do when you don't have a large budget, and the only awareness you can build for your business is through the *Yellow Pages*? The answer is that you have to be imaginative and look for answers outside the box. In his book, Dr. Glazier tells us how he overcame his business marketing challenges by reaching out to patients online, and he shares secrets he learned along the way that can enable businesses to float to the top of search results as an artifact of social media engagement. He shows us, with an amazing amount of data, how the old media, such as the *Yellow Pages*, are becoming obsolete, and he explains why. He demonstrates the power of social media and the ability to build communities to attract patients or consumers, And he shows us how, with the investment of just a little time every day, we can compete against bigger players and even become bigger players ourselves.

Social Media is a Force Few Understand and can Only be Understood if You Do it Yourself

You cannot benefit from knowing how to swim if you yourself never swim.

One of Dr. Glazier's messages is that certain social media tasks shouldn't be delegated, that there is a process of engagement and connection, from your thought leaders in your business to existing or potential customers, that should be a daily task. It is a mindset, one you need to be connected to your customers in this era of new media. It is one that stimulates you with new knowledge and power.

For corporations, this type of engagement is a culture change for the best, a 360-degree turn from process to innovation. It is an idea that engages you in thinking better and more creatively, from an intellectual, emotional, and visceral standpoint regarding your consumers. If you own or run a large corporation, getting your executives engaged in social media is the type of change you are looking for; it will make your team sharper and more focused, more in touch with the needs of your customer base and will make you more competitive in the marketplace. For corporations, the beauty of engaging in social media is turning your culture from passive to active; from one waiting for information and orders to one seeking information and taking responsibility at all levels; from a 9-to-5 culture to one that is engaged in the process of bettering the brand through connecting with customers. From couch potatoes to hunters and innovators, social media activists are those new executives with a passion for connecting emotionally with people outside of old, traditional, third-party media ... and the enlightened are the guys who will have a lot more fun and success as Dr. Glazier's entrepreneurial spirit rubs off on them.

Taking the Fear out of the Process

For many, social media is intimidating, and many executives don't yet understand, from a business perspective, what it potentially offers, or what kind of an "edge" executives can obtain by engaging in it. Still, better understanding the consumer, in order to increase brand competitiveness, and even entering into a dialogue with them that might lead to breakthrough ideas is no laughing matter for marketers large or small. This is the beauty of Dr. Glazier's book: it likely contains the type of ideas and answers marketers are searching for. His book takes you through the process of building communities online; the dos and don'ts of social marketing; opportunities to leverage social media campaigns to increase rank in search engines; and mistakes to avoid in the process. Dr. Glazier describes a "grassroots" social media campaign that can be scaled in unimaginable ways.

Searchial Marketing is a source of inspiration, a challenge for marketers to engage in social interaction, but most importantly, proof that these methods work. It is provided by a man who has implemented it and succeeded in eliminating traditional marketing methods while drawing more new business than ever before. I encourage marketers and designers, branding professionals, and executives from multi-national corporations to read and comprehend the passion and spirit behind social media as practiced by Dr. Glazier. Why? Because at the end of the day, social media is not a method or a process, a technology platform, or a new hyped media, it revolves around human engagement.

If you are a small business ready to move up, this book is even more important. It will save you money and time. *Searchial Marketing* will show you how it is now possible to leverage social media to create awareness and, ultimately, loyalty. If you are a large business, the message is even more pertinent, as corporations more and more will be defined not only by their products or communication but by the direct emotional engagement executives will have with their audience. It is critical to see, through Dr. Glazier's book, how one individual, in a quest for success against all odds, by taking steps to reach out to his patients and customers was able to build an emotional connection with people that no amount of money in traditional media could ever have bought. His sincerity and authenticity speaks volumes about the need for corporations to move out of their ivory towers, and for executives to spend more time with their consumers, ready to listen. The book shows that social media is not just another tool in your marketing arsenal but the core center of your strategy, the essence of who you are, and the spirit that needs to be conveyed across all other media. It is the emotional source and vital part of a brand essence. The book is challenging but inspiring, and it goes beyond traditional marketing as we know it to explore the extraordinary new channels now at anyone's fingertips. Learn how to effectively engage your brand with people in an unforgettable way.

Marc Gobe
President, Emotional Branding
Author, *Emotional Branding*

FOREWORD

I'm about to teach you things about your marketing methods you really don't want to hear. I'm going to show you that traditional marketing methods are relatively ineffective compared to newer methods; how traditional marketing methods cost thousands of dollars a month to implement with relatively low return on investment (ROI); and that traditional marketing is perceived by many to be unprofessional, thus traditional marketing efforts may be harming your reputation and that of your business or organization. I will also explain why it is difficult to compete with major companies that have large marketing budgets via traditional advertising efforts.

If I were to tell you there is another way to market—a more modern, more effective strategy that can generate new business, enhance instead of hurt your reputation, is extremely low cost, and potentially more effective in drawing new business—which marketing strategy would you pursue? Okay, you're not sold.... What if I were to tell you that by continuing to use traditional marketing, your efforts are bearing less and less fruit each year, and by not implementing the new methods, you will eventually have that much harder a time competing within this new marketing world? Not enough, you say? Well, what if I were to tell you your competitors will be off to a gallop with the new methods before you are even out of the barn?

I remember learning about Facebook in early 2007 from a friend. Frankly, it sounded like another way to waste time on my computer, and I was already rather proficient at that. In fact, I really didn't want to get involved in anything that cut into the time I was wasting on those things I was already wasting time on. I eventually created a free Facebook account, only because pretending to be into pop culture stuff makes me feel younger. After that first week of finding "friends" and being "found," I was hooked. When I was finally able to wrap my head around the concept, the potential of using Facebook to build business relationships simply overwhelmed me. Now I'm using tools like "Twitter," "stumbleupon," and other social media tools to improve my ranking in search engines—people searching for the keywords and key phrases that describe what I do and sell—as my sole means of marketing. Parallel to that, I am having tremendous success building relationships with existing clients while generating more new business than traditional marketing and advertising, and I am doing all this with my smallest marketing budget ever.

"Searchial" is a phrase I developed to describe an Internet marketing force that is growing and gaining in importance "Searchial describes the intersection of the social internet and search engine optimization – it is a word that describes how participating in the social interent causes the content you place online (your blog, website, tweets etc) to "float" higher in searches for the keywords and keyphrases you write about, enabling the people who are searching for the products and services you offer to find you better. I discovered and coined the term searchial when I realized my participation in social media was driving *new* patients into my medical practice as Google and other search engines were "favoring" the content I posted ahead of that of my competitors. Sure, I use it to build and strengthen existing relationships, but the most compelling aspect for a young, growing business is the aspect of driving new business just by taking part in this social Internet.

Twenty five and more years ago, marketing in many industries was considered unprofessional and rarely practiced. Only relatively recently have professional businesses started applying traditional marketing. Doctors began increasing their presence in phone directories like *Yellow Pages* only thirty years ago; a few even dare use direct mail, TV, and radio advertising; pharmaceutical companies use commercials on radio and TV; and even hospitals got into the game. The return on these methods is estimated to be less than 2 percent, and 2 percent is, in my opinion, pathetically considered an incredibly good return on the relatively large number of advertising dollars you need to spend to get such a poor ROI.

In the early version of the Internet, your organization needed a website to market. You set up a static page or several pages under one URL and waited for people to find you. Subsequently, you might have discovered and used paid Internet advertising methods, such as banner ads, pay-to-list indexing sites, and pay-per-click advertising offered by Google, Yahoo!, and others. This represented a new direction in advertising and marketing for the average business. Even so, very few who discovered these methods were brave enough to discontinue the old "tried and true" methods, and even fewer considered eliminating the *Yellow Pages* ad, which likely had the lowest value per marketing dollar spent. At that point, your marketing budget probably paid for traditional direct mail, print advertising, *and* Internet advertising; but again, the new clients trickled in, and you found yourself spending more on marketing and advertising than ever.

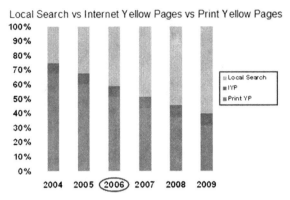

http://www.localsearchconsulting.org/improve-your-online-visibility/)
credit to http://sekinternetmarketing.com

Since you're reading this book, chances are you are implementing or plan to implement a social media strategy and are trying to learn as much as you can about social media, because of the *huge* buzz around the stuff. There is a fascinating new landscape unfolding, and an intersection in the landscape where social actions are recognized, even favored by search engines. Social media, which began as a communication tool, is now a very important business tool to be used within a larger, more compelling "searchial" marketing strategy. I will explain to you the value you can bring using social tools to drive new business into your organization not only through communicating and relationship building, but by improving your listing in search engines to rank higher. This way, searches done utilizing keywords and key phrases enable people to find the products and services you offer in your geographical area or even worldwide.

This book can be applied to and used as a guide for any industry new media marketing on the Internet at any stage of the game. For those of you interested in information on searchial marketing specifically geared toward healthcare clinics, hospitals, or pharmaceutical companies, the appendices are for you!

Dr. Alan Glazier

August, 2010

CHAPTER 1:
THE SOCIAL WEB; STARTING YOUR CONVERSATION

Social Media is pervasive in society and growing at a tremendous clip, surpassing other forms of electronic communication. Li and Bernoff, in their popular book on social media, *Groundswell: Winning in a World Transformed by Social Technologies,* state, "Media isn't neatly boxed into little rectangles called newspapers, magazines, and TV sets anymore. The Internet is not some sandbox that can be walled off anymore—it is fully integrated into all elements of business and society. People connect with other people and draw power from other people, especially in crowds."

Using social media, you can take advantage of this new "groundswell"; your message can reach thousands of eyes and ears with news and information regularly at the touch of a button! How is that different from e-mail marketing? When you use e-mail, you are pushing often-unwanted information at people. Social media conversations are two-way, where people who want to hear from you "connect" or "friend" you, or sign up to read what you write, so they no longer have to search and pull the content they are interested in from the Internet. Instead, they are giving you permission to share those things they want to read about. If the information you provide is quality, and relevant to the information your connections are asking for, your network grows and you reach more eyes and ears with your message. If your clients visit your business on average once a year, social media keeps your business in front of them through multiple exposures year round, possibly driving them through your doors more than once a year. While people frequently change their e-mail addresses, they are less likely to change social media accounts, so your lists are more accurate over longer periods of time. In 2009, Boston College stopped issuing students e-mail accounts in favor of communication via social media tools. I know what you're thinking: "But I just got comfortable using e-mail, and now things are changing again!" Well, all I can tell you is, buckle your seatbelt—you're in for quite a ride.

I'm sure at least half of my unsolicited patient mailings and other "push" marketing efforts were glanced at and trashed after I paid a fortune in printing and postage. At best, each contact I was able to reach through traditional marketing had, on average, a reach of a family of four, and maybe a few friends. I question whether anyone even *looked* at my *Yellow Pages* ad. Now when I have an interesting comment (affectionately known as a "tweet") to post on Twitter, it can get proliferated ("re-tweeted," or RTd) to hundreds of thousands, even millions of people! And each new tweet is recognized by Google and appears in the top of search engine searches for the keywords and key phrases I input in the tweet. When I finally understood how to manipulate this to my advantage, I wondered what reason there was *not* to be part of this new "conversation." It was at this point that my social media efforts took off, and within a year, I had built a significant network of thousands of "followers" over several social media outlets.

Today, my practice marketing budget is reduced more than 80 percent, the "unique new visitors" statistic for my practice website is up more than 100 percent, and new business continues to roll in to the practice without using any type of traditional media advertising. My website, which after fourteen years receives thousands of visitors a month, has been eclipsed by my "blog" (more on blogs and blogging later), which within one year of launching receives seven to ten thousand visitors a month. I have enhanced my reputation and that of my business, enhanced my visibility, and great opportunities are coming out of the woodwork, such as professional blogging opportunities, interviews by major industry and non-industry publications, and regional awards.

You are probably thinking, "Yeah, great for you, but I don't have the time." That may be the case, but with a small bit of patience, even a surprisingly smaller effort and some decent *delegating* skills (the key to maintaining your sanity) can help you use social media effectively to market your business while reducing the expense and effort of traditional push marketing (e.g., direct mail, recall (contacting clients to remind them about annual or recurring visits), *Yellow Pages*, etc.) and *profit* from it.

Don't worry—you don't need all the intimidating tools like Twitter and Facebook to start. We will dip our toes in the water one tool at a time and build your conversation from the ground up—the best way to start a successful, long-term effort. There are a few rules for success in *Searchial Marketing* that I ask you to pay close attention to when participating in searchial marketing (SM). Once you immerse yourself in it, the differences between the people who follow the rules and those who don't will become apparent; those who follow the rules benefit with increased business and, opportunities and an improved reputation, obtained through their growing network. Those who don't are ignored and never maximize the potential of the campaign.

Rule #1—This is a Conversation—Participate

It's not enough to set up your LinkedIn profile, Facebook page, or Twitter account. Don't bother if you plan on having it remain static—you already have a website for that. The purpose of participating in SM is to interact with others, establish a rapport with a wide variety and network of people, and create interesting posts and content that draw people to your efforts. This only happens if you are participatory; there are plenty of voyeurs out there watching the conversation (don't ask me why), but those who get the attention are the ones who are social within—hence the name "social" media!

Rule #2—Pay Attention to the Conversation

You'll find you need to keep up with a lot when you participate in more than two SM "sites." (facebook, Digg, Twitter etc) It can seem overwhelming. I felt much the way I did at the craps table before I understood the game—I didn't want to play because it seemed intimidating, but there are really only a few things you should do consistently to maximize your odds of success while ignoring the useless junk (or "sucker bets," as they are called). You will soon be able to focus only on what counts, and you will ultimately use multiple tools to promote the content you create. Don't be afraid to get heavily involved in multiple suites of social media now, with the ultimate intention of "pruning the tree"—eliminating the SM suites that you don't derive value out of as your efforts progress. You'll need to watch others, and pay attention to how they interact in their "conversations" on the different suites so you can learn how to build your social skills, learn the lingo, and rise to the top like cream without sticking out like a sore thumb.

Rule #3—Don't be a Walking Billboard

People don't want to be pelted by advertisements or posts that may be interpreted as advertising. You will lose followers left and right if you do this, and your efforts will amount to nothing. If you have a product, and you want to use social media to move it, you have to be subtle. If you are selling a line of antioxidants or other vitamins, start a blog on health, create useful content, and hide your plug within the content. Don't shout in a tweet, "Buy these special antioxidant pills—click this link." People want to friend you or follow you because your posts are interesting, substantive, and bring value to the conversation; they can and get pitched everywhere else, but you can be the one grabbing their attention and making meaningful connections.

Rule #4—Get Ready to Give Five Times before you Ask Once

You want people to follow or connect with you when you participate in SM. A "follower" is someone who "signs up" for your content feeds and in doing so allows you to push information at him or her when you want to. You control the content of the information that you push—just think of the possibilities. People tend to follow you when they see you are taking the effort and spending the time to create content valuable to them. Your followers are basically giving you a license to push information to them as much as you want. As long as you provide quality information, you maintain the majority of your network and attract new followers as well. It's okay every once in a while to promote something, just do so sparingly and maintain professionalism when you do, and you'll see your network thrive and grow. It's okay to plug something 20 percent of the time, but you need to give something of quality to the conversation 80 percent of the time for people to accept an occasional marketing message from you.

In the early days of the Internet, websites were used to market businesses. Marketing using websites in the early web was called "pull" marketing—you put your website up and waited for people to find it. Your website was static; there was nothing interactive on it, and it sat out there in cyberspace, waiting to be found. You could spend money to pull people to it, just like you still can use banner advertisements or paid directories. Basically, you put your website up and waited, or spent lots of money on advertising and waited. If you needed to change something on your website, you either had to learn computer languages or spend decent money on a developer. That version of the Internet, referred to as web 1.0, is obsolete.

We are now well into the third version of the Internet, Web 3.0, where "push" is the new game in town. Push marketing involves creating content (sharing knowledge and expertise through blogs, video, or audio means), and people who like the content you are creating or what you are doing "sign up" to participate in a conversation with you. They do this on social media websites by connecting with you either by invitation or by finding you based on the subject matter you are creating or proliferating and they are searching for. Connecting with you is an indication they are interested in what you are saying. When others connects with you, they are essentially giving you permission to push your information toward them. Web 3.0 is dynamic and full of interaction and socialization. A company's web presence can be regularly updated without learning computer language or spending a fortune on developers via a blog or social media platform/site. In Web 3.0, you don't have to pay to advertise; you have to put some work into it, in the form of creating ongoing new information relevant to the subject matter your potential customers want. Pay per click still exists, but people participating in Web 3.0 are more likely to trust a highly ranked "organic"—i.e., naturally occurring—search engine listing than an ad they know is sponsored to push information. They are also more willing to trust someone who has been "recommended" online by someone else, or to look up the person or businesses "influence" as ranked by social influence-ranking websites such as Klout.com. If you want to pay, you can use pay marketing services that enhance your efforts even more, but you don't have to have an expensive marketing effort to be successful in this realm. Most of it is intuitive, and almost all of it is free.

Web 3.0 is ultra-dynamic, with exciting things like "augmented reality" (already in its infancy), location-based advertising, and information gleaned from modern bar codes, called "QR codes," which are read through the camera on your mobile phone via a mobile application. Imagine walking down a street in New York City and holding your PDA in front of you while you walk. On your screen is a virtual picture of the street and buildings in front of you. In cartoon fashion, bubbles with advertising and other promotions pop out of the virtual storefronts whose brick-and-mortar counterparts are steps ahead of you. Or you hold your phone up at the entrance to a hospital, and a floor-by-floor map shows you exactly where radiology is. While shopping, you hold your phone camera up to that product, information appears on your screen, and you get a coupon for the product on the spot. You use this information to learn things you otherwise wouldn't know about: sales and promotions, etc., and stores use it to draw new business right off the street. Global positioning systems (GPS) are used to shoot messages to you or for you to locate and find friends, or research restaurants and professional services like doctors' offices. Google's Android technology is leading the way into this strange new world of Web 3.0.

The Internet continues to evolve. Now, businesses that use *Yellow Pages* are considered geriatric while your younger clients (age fifty and younger—which should make most of my readers happy to be included in the younger group) are searching for products and services using Internet search engines. Direct mail goes in the garbage and the *Yellow Pages* is used as a booster seat for your three-year-old child. (It is rumored that the postal service is considering ceasing operations, as "snail mail" keeps getting more expensive and is totally inefficient compared to other methods of communicating, such as e-mail and social media.) Need a neurologist? Google one. Looking for the closest pharmacy? Google one. Searching for information on a healthcare corporation? Ask a question on Quora.com...

The first thing you should do is drop direct-mail ad campaigns. I suggest you try dropping one traditional marketing method every three months for a more comfortable, less stressful transition as you ramp up your social media campaign and search engine position. "What about the *Yellow Pages*? Can I really drop that too? What if someone can't find me?" That's where you need to adjust your thinking cap. It's a new marketing world out there I know from experience; I reluctantly did what the management experts in my field recommend and dropped my major *Yellow Pages* advertising at the end of 2007, and I remember what it felt like ... but no one is finding you there, so ask yourself if the population still using the *Yellow Pages* and things like mailbox flyers is the population you want to attract to your business. So why continue? The significant majority of people with disposable income search for the services and products they seek on the Internet using search engines. Wouldn't it make more sense to spend your *Yellow Pages* money in areas people are searching? See where I'm going with this?

Privacy

You will have to "put yourself out there" if you want to participate in social media, plain and simple. Are there privacy concerns for social media? Absolutely. Most privacy concerns, however, are overblown. Can you avoid exposing yourself? You can try. There are things you do and don't want to do when you participate. Many social media platforms allow you to maintain multiple sites, and you can separate your business from your personal use. For example, Facebook makes it easy for you to separate your business page from personal. You can even set preferences that keep your business from finding your personal and vice versa. Privacy settings are important if you are worried about privacy, and most social media suites offer them. Tip: if you are concerned about privacy, don't go around the Internet clicking "like" buttons. Every like button you click collects information about you. Privacy is a legitimate concern.

There are three kinds of people surfing the Internet: the overly cautious, those with a come-as-they-may attitude who deal with privacy issues if and when they arrive and those who hover somewhere in between the two. You can bank manually, you can avoid e-mail for fear of getting a virus, and you can avoid filling out online forms for fear of identity theft. You can do tons of things that will limit your exposure and reduce risk, but at the risk of less exposure for your business. Remember, social media doesn't have the word "social" in it for nothing. Maximizing your exposure is a key to your success in social marketing. I'm not advocating that you do things you are uncomfortable with or that put you or your business at risk. Most of social media involves people interacting, much like a cocktail party, so you can hide behind the punchbowl, but you won't make connections and are sure to not be invited when something exciting is happening.

Time Constraints

For a social media campaign to be successful, you or someone in your organization will have to invest time and effort. The most time and effort is expended getting your campaign off the ground, and moving forward creating content becomes the largest and most important time expenditure. You will create, post, delete, correct your mistakes, repost, learn, and adjust, and your campaign will grow. It can take three to six months of diligent work before you start to see significant rewards. As time goes on, you will learn which tools work for you and which don't. You prune the tree of social media tools down to a few effective ones, and you find tools that help integrate those tools together, minimizing the amount of time you spend posting. Technology changes, new tools surface, and you spend time learning to implement them and analyze whether they fit within your strategy. When they do, you move older tools out to implement newer tools.

The footprint of your social media efforts evolves in this manner. At some point, the work gets sophisticated, and you will likely have to delegate, maybe even bring in a marketing specialist to handle the new media for your organization. The two keys to limiting the amount of time you spend on social media are proper tool usage and delegation. While there are some things you may be comfortable delegating, there are other things you may not be. If you are like me, I don't want people writing for me; everything written needs to emanate directly from me. I don't recommend delegating creation of content, but I do require my staff to come up with ideas for my blogs. I discovered my inner writer by creating content for my social media efforts. If you are certain you don't like to write, you should delegate the writing. I don't know of a successful social media campaign that doesn't revolve around original content. Be sure to involve yourself as editor with final copy approval rights, so you can maintain control of your content. Ask your staff to write posts; they often provide unique and interesting perspectives. You can ask vendors, other professionals, and even customers to be a guest writer/poster on your blog. You don't have to write a post to post it—find a useful source of updated information in your specialty, copy the article into your blog (giving appropriate credit to the authors), and draw attention to it to through your social media contacts. All of this can save you time and energy, but content is king in *searchial* marketing, so if you don't want to write, you need another way of finding fresh content, including hiring a ghost writer.

How much Time should I Spend on Social Media?

At first, you might spend a lot of time practicing and learning to use the tools you need to succeed. You might spend some more time creating content, such as blog posts, or trying to build your networks, but after the initial push (anywhere from three to six months), you can expect to spend less than an hour every other day working on this. I post between three and nine blog posts a week, communicate with my networks between patients or when I have time due to a cancellation or no-show, and usually put in a half hour in the morning before patients arrive at my office. I use tools that automatically post between my live posting, blogging, and microblogging (more on blogging and microblogging later). I might write a week's worth of posts on a Sunday night and publish them over the next week. Sound like a lot to write? Probably, and you don't have to spend as much time as I do to run a successful campaign, but my chief motivator is my extremely low marketing budget. By shifting my time and participating in social media, I save more than $80,000 per year in direct marketing expenses and have a much greater response than I ever had with traditional marketing. You can expect to get out of social media what you put in, so dip your toes in the water slowly, work within the time constraints your are comfortable with, and as you achieve success, watch your business grow and your expenses decrease. Six months later, ask yourself if the results warrant more of your time, and if not, stick to the plan or delegate so you don't feel you are spending too much time on it.

Hutch Carpenter is a seasoned blogger, guest blogger, and author of *I'm Not Actually a Geek*. His observations on technology and business—as "someone who should know better"—shows on his graph (see below) in a guest post that he defines as "stages" of blogging. In the graph, he demonstrates the "ramp up" a blogger will experience in terms of content creation during the early and middle stages of the process. As the blog grows in popularity, the audience expects the blogger to produce higher quality posts. Thus, the trend tends to be that volume of content produced suffers as quality of content increases. You should expect, eventually, to need to produce less content over time, but what your audience will expect you to produce will be more original, thought-provoking, high-quality content.

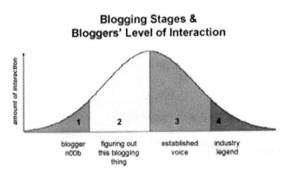

Posted on http://bhc3.wordpress.com/2008/07/
JULY 22, 2008, Hutch Carpenter.)

The evolution of your marketing and advertising efforts will naturally shift toward using new media whether you want it to or not. Delaying the inevitable will harm your organization. Working with new media to market your business will produce results that go straight to your bottom line and cut your advertising budget as close to nil as you want to go. This is the story of how the social Internet is capable of driving customers into your business and how progressive businesses are leveraging it to find new business and decrease marketing expenses. It's gearing your social efforts toward search engine optimization, or, as I call it, searchial marketing.

Within today's Internet, there are four broad methods you might use to market a business. The first but not necessarily the best method is paid advertising, including banners, pay-per-click, paid directory listings, and a few others. The second category is utilizing social media to help improve search engine position (for which I coined the term *searchial*), for the most part, without significant expense. Third, you can use your networks to leverage your business. Fourth, you can do what I call grass-roots social media marketing, in which you collect existing and new clients within a group or page and socialize while indirectly marketing your products or services. The searchial method provides the most return on investment; by interacting in the grass-roots social space in a calculated manner, you cause your web content to float to the top of searches, driving in new business.

Costs you may encounter using searchial marketing are mainly development expenses (e.g., hiring website html and design experts, or consultants who can structure the code for such things as the website content, blog content, and attributes) behind your content. Once you have the code and design in place, you produce and add content peppered with keywords and key phrases people use to search for the products and services you offer. The majority of your searchial efforts can come at minimal cost by using many of the methods and tools I will share with you throughout this book.

Paid Advertising Methods in New Media

Pay-Per-Click Marketing

Not considered a new form of advertising anymore, pay-per-click or impression advertising describes advertising services offered, typically by search engine companies like Google, Yahoo!, and Bing, that, for a fee, put your advertisement in front of web surfers who input a specific keyword or key phrase. Keywords are words humans input in a search tool, such as Google, to find information on that particular word, and key phrases are phrases that are used for the same purpose. The ad usually appears grouped with similarly targeted ads and runs along the right side of the browser page or across the top. Fees paid for these types of ads are based on what the market will bear, meaning that if your competitor is willing to pay $.25 for each potential customer that clicks on their ad, you will have to pay $.26 or more to show up ahead of your competitor. There are different ways the fees you pay might be structured.

Click Through Rate (CTR)

An "impression" is registered every time your ad is shown in any browser anywhere. A "click" is registered when someone clicks on an ad. Pay-per-click and cost-per-click (CPC) fees are calculated on the number of clicks that take a visitor through the link you provide divided by the number of impressions. You specify the amount you are willing to spend each time a visitor clicks on your ad. This method is used to have better control over the amount you spend on this method of advertising. If you are attempting to gain direct sales or visits, the CPC method is the way to go.

Pay for Views

Pay for views (also called cost-per-thousand impressions or CPM) tells how much you are willing to pay each time the amount of visitors who click through your ad total one thousand views. The advantage of CPM is greater exposure, and people who use this method are generally less concerned with how many direct clicks it generates as they are trying to get clicks *and* exposure to help build their brand. If you are marketing your services within a defined region, CPM is a better option as while you may not have as many direct click-throughs, you are more likely to expose your brand to more people in your vicinity.

An innovation brought forth by Google recently is the "cost per action" (CPA) rate for which you pay. No more pay per click or impression; you pay when someone completes a transaction! This is a risk-free way to reach a larger audience.

Most CPC/CPM campaigns have a minimum cost per click of $.01 and a minimum CPM of $.02. While Facebook says that all ads compete equally for placement regardless of how much of a CPC or CPM you are willing to spend, it is generally believed that the higher the CPC or CPM, the more frequently your ad will be shown. For most CPC and CPM campaigns, you set a daily budget, and the system ensures your daily charges will not exceed what you state as your maximum. Many sites, such as Google and Facebook, have a bid estimator that makes it easy to see what average charges you can expect daily based on the ad or keywords you are targeting with your ad. These estimators are extremely helpful when trying to stick to a budget. Once your budget is reached, your ad is no longer shown during the remainder of the day. You can change your daily budget any time, and the new charges become effective within an hour of your changes, in most cases. No site will provide any guarantee as to how many clicks you will receive.

On Facebook, the minimum daily budget for CPC and CPM is $1.00 USD. In addition, your budget must be at least two times the CPC or CPM you have specified. For example, if you specify a $10 CPC, then your daily budget must be at least $20. For Google Adwords (more on this later), the minimum cost per click is $.01, and the minimum cost per thousand impressions is $.25. There is no minimum budget limit.

All CPC and CPM ads are shown in similar positions on most search engine and social media sites, so your ad and your competitors' ads will show up in similar positions on the search page, but at different times. On Facebook, it is usually on the top right-hand side of the page. Google Adwords shows ads down the right-hand side of the page and across the top.

Google has another advertising service, Adsense. Adsense enables other business ads to show on your web page or blog page. Adsense is described by Google as, "By providing ads tailored to your interests, we offer useful tools for you to view and manage the information that is being collected and used to serve ads. Google AdSense is a free program that empowers online publishers to earn revenue by displaying relevant ads on a wide variety of online content." It can be a good way to monetize the content you create.

Most sites set a limit on the daily amount you are allowed to spend. Most sites will increase your maximum limit as you prove credit worthiness. The longer you advertise on the site, the more they will allow as a daily maximum.

Facebook answers the questions, "Why is my average cost-per-click or CPM less than my maximum cost-per-click or CPM? How does Facebook determine my cost-per-click or my CPM?" with the following answer posted on their site:

"For any given ad unit, we select the best ad to run based on the cost-per-click or CPM impressions and ad performance. We have a process in place that will automatically calculate the minimum price that the advertiser could pay and still have the highest cost-per-click or CPM ad, and the advertiser will only be billed that price. This price may be below the advertiser's maximum cost-per-click or CPM. Because we lower the cost-per-click or CPM on your behalf, we recommend that you enter your true maximum cost-per-click or CPM when creating an ad. This will increase the likelihood that you do not miss out on clicks or impressions that you otherwise could have received."

Social Media Marketing; Grassroots Effort

For someone newly entering the realm of SM, it is logical to assume one-to-one connections would be the way people would expect to drive customers to their business. Maybe it goes something like this: you put yourself out there, someone finds you, that person visits your location and has the best experience ever, and he or she refers twenty friends after you give that person the best service of his or her life. While this happens occasionally on a micro-level, these business-to-consumer efforts don't happen frequently enough to make them the main thrust of a campaign. You won't generate much business communicating directly with people via text, e-mail, message board forums, or social media interfaces one-on-one. You might eventually convince them to patronize your business, or expose them to it so many times it generates a new client that makes social media campaigns most effective. Even if you were able to achieve this, there wouldn't be enough time in a day to make it worthwhile from a business standpoint. Social media campaigns work by building relationships and repeatedly exposing your online brand to the people in your network. This type of social networking doesn't send a ton of new business rushing through your doors, but they help *some* new business find you. They are *mostly* geared toward *increasing* business from those who already patronize you.

I call this method of social marketing the grassroots effort. Grass roots describes communicating with groups of "friends" or "connections" on social-media sites like Facebook and Twitter with the purpose of building relationships and staying in the "eye" of the network. The most popular grassroots Facebook users, tweeters, and bloggers are those who use their particular expertise to create useful content and post it within the social web. People become your connection, or your "friend," because you bring something of quality and value to the conversation. What you bring to the conversation is usually your particular expertise in the form of content you create. Your connections increase over time when you produce quality content. People connected to you will re-broadcast content you have posted when it is compelling to them and they think their connections will benefit from coming in contact with it, driving more people to your efforts. Some of those communications will result in new customers, but most of this effort is geared toward growing a following enhancing your reputation in the eyes of your followers, and providing repeated exposure to those in your network who already patronize your business. Using this method enables you to open up a direct customer-communication channel, inviting discussion and feedback with existing customers. Social media is an excellent customer-service tool and can help maintain and build on existing relationships.

Grassroots social-media efforts are time consuming and require a lot of attention. They involve creating content, responding to messages and reputation monitoring on the Internet. Delegation is key to a successful grassroots effort; the time and effort it takes to stay on top of client interaction can be daunting. Assign one or two social media sites per staffer and have them responsible for maintaining it. I like to control the content that I put out there, so I will direct my staff to finished content and task them with the duty of proliferating it to our network, or copying and circulating pertinent information based on specific patient questions. Responding to customers should be done with kid gloves and involve phrases like, "Have we met your expectations?" and "How can we improve on the services you receive?" In a grassroots effort, your business should ask questions that help you collect data on the services you provide, enabling you to figure out what needs to be done to make your business a better place for clients to visit. Sending a message like this can go a long way toward making a client happy, or making someone seeking your services feel particularly well cared for.

You can have a thousand people "like" your content, and a network of tens of thousands of people you attend to via the grassroots method, but if those people aren't in your geographical vicinity, and you aren't selling a product over the Internet, business isn't walking in your door. The network can, however, create a "buzz" around your content, which someone in your vicinity may pick up on at some point, generating new business for you through the grassroots method. The searchial method, in addition to grassroots social-media campaigns, can enhance your return on investment in social media much greater than grassroots social media alone.

Searchial Marketing: Where Social Meets Search

A well-known optometric practice management expert put the bug in my head: you're losing money with *Yellow Pages* ads. I hated to hear it, maybe because deep down I knew it was true, but I was too scared to let go. Optometric practice management experts have generally given me good advice, and when I've followed it, more often than not it's helped, but this was a big step. Was I ready? The idea scared me, but after a year or more of the idea resonating in my head like an old iron bell, I took the plunge and scaled back my *Yellow Pages* ad, taking my cost from $1700 per month to a mere $400 per month, leaving me with single-line listings for each doctor at my practice.

I never did direct mail, and I don't use practice newsletters, so the growth of my practice was dependent on internal marketing efforts and word-of-mouth referrals. After I eliminated my half-page *Yellow Pages* ads, I was left feeling totally exposed. How were new patients going to find me? Would anyone even know I existed? The thought alone was nerve wracking. Not being someone who sits around and waits for something to happen, I started looking at what my advertising options were. My goal: to drive new business into the practice without engaging in less-than-professional marketing tactics. I read marketing books and consulted with experts. I found that most of my patients were searching for healthcare through the Internet, and I started an expensive Google Adwords campaign. In late 2006, a patient of mine, Shashi Bellamkonda, the social media "swami" at Network Solutions introduced me to a marketing phenomenon in its infancy, which was just beginning to thrive within the new Internet, a social media aspect of Web 2.0. "Search engine optimization" (SEO) is the process of improving the visibility of a website or web page in search engines via "natural" or unpaid search results. I think a better term to describe this tool is "search engine elevation." How you are "optimized," or "elevated" through any particular search is due, in large part, to the content you produce and proliferate on the Internet. When I realized what this could do in terms of driving new business through the doors of my brick-and-mortar business, the lightbulb went off, and I decided that, instead of spending money on traditional marketing, which was becoming less and less effective as people searched more and more in search engines, I would try to master this new media channel.

Dr. Alan Glazier

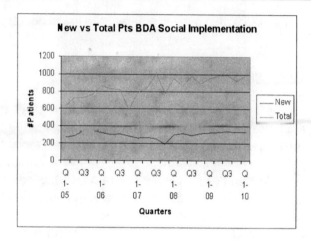

This graph shows how my practice fared, from the time we used traditional marketing and paid Internet advertising (using adwords and directory advertising), through the time we shifted our effort totally to searchial marketing. BDA stands for "before, during, and after."

From 2005 through the middle of the first quarter (Q1) of 2006, we had about $1700 per month of expenses on *Yellow Pages*-type of advertising, another $1000 per month in community directory advertising, between $1500 and $2000 per month in Adwords advertising, and about $1000 per month on pay-advertising medical websites. Total number of patients during this period experienced significant growth, but sometime during the middle of the fourth quarter (Q4) of 2005, the volume of new patients started to slip.

During the second quarter (Q2) of 2006, through the middle of Q4, there was a significant decline in total patients, and a small–but-steady decline in new patients that continued through Q1 of 2008. We cancelled our *Yellow Pages* advertising effective Q1 '08, discontinued all community directory advertising and pay-advertising on medical websites in Q1 of '08, and began our social media and search elevation efforts, mostly through Facebook at that time.

This was where things got interesting. For the first time in nine quarters, we realized an increase in new patients, and it was a dramatic one. The increase was timed precisely with the *cancellation* of most of our print advertising and the discontinuation of our adwords campaign. Historically, we had witnessed periods of business growth where total patients were increasing, yet new patients were not, and I believe this was due to the fact that we placed a lot of marketing emphasis on existing patients, utilizing internal marketing efforts that improved customer service, an overhaul of our recall efforts, and a patient-friendly change in staff and policy.

While our efforts saved the slide and apparently turned around our hard work to attract new patients, they also cut our marketing costs by about 85 percent, possibly more. Our internal marketing changes, plus our social media implementation, have resulted in incredibly healthy practice growth. As our social-media efforts expanded, and we improved them, I expected to see a dramatic up-turn in new patient numbers, and our strategy was to retain more of those patients via our internal marketing efforts.

So How Does This Work

Since you are reading this book, you likely have a website. This makes it likely you have content you wrote that exists on the Internet. Search engines have software, referred to as "bots," or "spiders," that are "sent out" across the web regularly. The job of these bots is to "read" content and assign levels of importance, or "relevance," to websites and blogs based on keywords and key phrases found in the content. For instance, if your content regularly mentions the keyword "optical," the bots award you a certain numerical value that equates to the "importance" for searches that include the keyword optical. If you have the word "optical" linked to other sites that mention the keyword "optical" and they link to you, then your content earns a higher numerical value. Thus, your content is assigned a higher importance score for your content than sites that do not. Google uses the term relevance to describe the importance of content for particular keywords or key phrases. The relevance of a website and/or blog is determined by many things, but mostly by how much up-to-date relevant content is published in a given area of expertise and how many other websites, blogs, forums, and directories reference the blog or website by linking their content to it. When a bot finds a site that:

- Is regularly updated with content that is keyword-specific to a given search;
- Links to other highly relevant sites within the same specialty;
- Is modern in its technology (incorporates online forms, videos, etc.); and
- Participates in certain social media efforts,

the bot then assigns a higher grade to that site or blog than it would to a similar site in a similar area of expertise that doesn't have these qualities. This higher grade helps the content get placed higher in the search engine listings when an Internet user searches for the keyword the bot recognized and rewarded it for. In essence, the more "relevant" your website for any specific search term, the higher up you will appear in search engines when people in your geographic location search for your services. As stated earlier, I refer to the process of modifying content to improve a search as search engine elevation. Eventually, you show up higher on a Google search page for optical than your local competitors and will be "found" more by people in your area searching for optical products. This is the most direct value proposition for using social media for your business—getting found more frequently than your competitors when people are looking for the goods or services you offer. The bots can be your friends or your enemies and are extremely powerful; I sometimes refer to them as the bot gods. Respect the bot gods—you may even want to pray to them (I say jokingly)—because the bots show their love by increasing your position in search engines, and there is a direct correlation to where you are found in an organic search and business revenue.

Participating in the social internet by proliferating content increases your relevance and influence from the perspective of search engines for searches that includes specific subject matter of your content. As you publish and participate in "the conversation" on the social internet, your relevance as an expert increases, your influence online grows, and your content "floats" to the top of specific searches of keywords and key phrases included in the content you write. "Searchial" describes this new and evolving intersection of social media and search engine elevation as well as a way to "act" to ensure the success of your online marketing efforts.

Leveraging your Network

There is power in numbers; the larger the audience, the more the eyes to see and ears to listen to what you have to say. The more people listening to what you have to say, the more influence you have. More influence equals better bargaining power. This is how, as your social networks grow, your bargaining power with vendors increases. How do you take advantage of it? Do you leverage the power of multiple offices to decrease your cost of goods? Do you leverage it to decrease your print, ad, or radio marketing expenses? Are you part of a collective bargaining network where you pay a fee or a percentage of the gross or net of your business? Social media is no different. Spend the time and energy to build a large social network, and you will have a tool you can use for bargaining with your vendors. Remember *Field of Dreams*—"If you build it, they will come"? Not true with social media. Building a social network large enough to become a bargaining chip takes time, creativity, and interaction with the network; after all, it is *social* media. Scared or reluctant to put in the time and effort? Think about it—the time and effort it takes to build a social network is nothing like the time and effort it took to build your business. To keep your marketing efforts modern and effective, this is where you need to turn; since it will save you money, why not?

You can choose to follow one or more strategies, but each strategy takes time, and unless you can delegate a significant amount, you won't have the time to pursue both. Later in the book, I will discuss time-saving tools you can use in your social-media efforts.

CHAPTER 2:
THE "SOCIAL" SIDE OF SEARCHIAL

Relationship Building

I grew up in the middleclass town of Wheaton, Maryland. My grandparents lived about eight miles away. My grandfather was an apartment building owner and manager, and as a result, was quite handy. I remember trips to the local hardware store, Triangle Hardware, where he was served by the owner and an old guy behind the counter. They worked there for twenty-plus years and had built strong relationships with the community they served. I can still remember the narrow aisles, the bins full of nuts and bolts that I would run my fingers through and play with while I waited for my grandfather to finish shopping, and the smell of sawdust and oil that permeated the store. Memories, like playing with the levels and watching their bubbles go back and forth are still so vivid to me, thirty-five years later. This was in the 1970s, and who could foresee that in the next twenty years, the era of the mom-and-pop store would disappear into history, replaced by the era of "big–box" stores? Amazingly (and in my opinion, fortunately), small business is experiencing a rebirth, thanks to social technologies; social media is allowing the small business to compete in the world of big-box stores. I am not saying WalMart is going anywhere, and I am not waiting for Triangle Hardware to reopen, but the strength of the mom-and-pop store was relationships with customers like my grandfather, and there is a resurgence in the relationship aspect of business occurring in online commerce via social media. You can find mom-and-pop stores online. For those too young to have experienced businesses forged through personal relationships, they were strong relationships, the kind of relationships that set a strong bond between goods and service providers and trusted clients; clients who *refer*. Social media is a return to what the neighborhood restaurant or mom-and-pop business provided: trust, personalization, and willing dissemination of expertise. Business built on relationships.

Building the kinds of relationships you want through social media requires *listening*. Pay attention to your network. Mine your network to find out which subjects they want to read about, and take the time to create quality content in response to their needs. When they see you are writing what they want to read, trust grows, the relationship solidifies, and word-of-mouth expands your reach. *You* are responsible for making the connections of relevant web conversation to your business, and you do that by plugging yourself into the groups of people searching for the information you happen to be an expert in and can provide for them. Take the massive conversation occurring on the Internet, filter it down to the discussions relevant to your expertise, and steer the people to you and your staff, where you can engage those seeking your information and bring value to your business as a result.

Tom Peters, a management consultant, most famous for having co-authored *In Search of Excellence*, a seminal book on business management, uses the concept of Return On Investment (ROI), or to coin a new term relative to the social web, Return On Investment in Relationships (ROIR).

Here's a take on ROIR from Harry Markopolos, author of *No One Would Listen: A True Financial Thriller*: "The financial industry is a business of contacts and relationships. No one ever buys a product and says, 'That product is the sexiest thing I've ever seen. I don't care who's selling it.' Generally, people do business with people they trust and like, or people who are recommended by someone they trust."

Peters goes on to say, "Over the weekend, consider in detail your ROIR strategy for next week, the next month, maybe the rest of the year. This is an idea that deserves careful and continuous thought, not a catch-as-catch-can attitude. You'd work for months or years on a plan for a new bridge. Well, ROIR is your 'bridge' to success."

The relationship side of business is growing, and its growth is providing a new opportunity for businesses, to compete with big-box providers. This is happening largely as a result of the popularity of the social web and the new methods that make connecting with customers and building relationships, some of the more valuable aspects of business, more simple.

You will be social within the searchial Internet, and you will parlay that into new business, but more effectively than with social alone.

Getting Started

To start your searchial marketing efforts, you will need information: articles, videos, and books that introduce you to the social sphere and the tools that exist within. Fortunately for non techies, the most prolific content writers out there are writing about social marketing and social media. There is a slew of good information online, and I strongly suggest you get your hands on as much of it as possible before you start signing up on social media sites.

Mashable.com is a great place to scan the hottest topics and learn from others in social media, Web 3.0, and mobile technologies. Mashable is a blog that has more than eight million monthly pageviews. It ranks as one of the largest blogs on the Internet. If there are new ideas, trends, or buzz, it is likely it will be covered on mashable.com. Smashingmagazine.com focuses on trends and techniques in web design and development and is an excellent blog to visit once you are familiar with search engine elevation strategy, which we will dig into in the next few chapters. Readwriteweb.com is another excellent web technology blog with industry news, reviews, and analysis. It receives consistently high rankings by websites that rank industry blogs. SEOmoz.org is a particularly excellent source of free information, although the best information is obtained through a paid subscription. I do most of my learning through SEOmoz.

On Twitter, social media experts proliferate content on social media topics you might be interested in. I recommend using the site search.Twitter.com and typing in terms like "social media" or "search engine optimization" to see who and what pops up. If you like a particular author's content, Twitter makes it easy to "favorite" the post to follow for easy reference. You can also connect to social media leaders on Twitter. It is also helpful to find Facebook or LinkedIn groups that discuss these topics, and there are plenty of them. I would avoid anyone who calls himself a guru; there are too many people out there recreating themselves as social media experts and gurus, so be careful. All the information you need is available for free, so don't get sucked into fancy Internet video presentations where people are selling their "consulting expertise" regarding "thought leadership marketing" or "Internet marketing"; there are a lot of carpetbaggers out there. Don't buy fancy products or software that promise to get you millions of followers.

The only way to run an effective social-media campaign is to build the trust of your connections and followers, and that can only be done slowly and carefully.

Peers and colleagues likely have experience using social media within your particular area of professional and academic interest, be it medicine, business management, retail, communications, or human resources. I recommend searching your colleagues' social media feeds, blogs, and websites to see how they use social media. Mirroring is a good place to start, and then develop your own individual style, or use some of the coaching tips found throughout this book. Many organizations are involved in social media, and conferences, for example, use "tags"—called "hashtags," symbolized by the same pound sign on the face of your telephone. They use this on Twitter to organize and publicize their events. For example, a popular tag used to locate trending information on healthcare, combined with social media, is #hcsm. If you type this in the Twitter search bar, you will find thousands of tweets discussing social media and healthcare.

Most professions have Internet forums where information is exchanged freely between colleagues. A simple way to find useful forums is to use the "Forum function" within Google. In the Google search tool, type the phrase you are looking for, followed by + forum. For example, if I was searching for forums where optometrists exchanged information, I would type optometrist + forum in the Google search bar, and listings would come up where optometric forums exist, or where optometrists are mentioned in forums.

How the Game is Played—Rules:

1. **Create original content.**
2. *Glazier's Rule:* There is no limit to the amount of content you can create and publish, as long as it is has more value to your readers than to your business.
3. **Limit** the publishing of content that has more value to you than your connections. By practicing this simple rule, you will establish yourself as a thought leader and your networks reputation will grow as a source for the subject matter you post.
4. Incorporate popular keywords and key phrases that people use to search for the products and services you offer into your content.

5. Hire a developer and work with him to fix the proper tags and header codes to your website or blog.
6. House your content within your blog or website.
7. Drive people to your content by proliferating links across the Internet.

In the next part of this book, I will teach you how to use social media to increase your search engine ranking, which will drive new customers to your business. While you are helping your business grow, you will be improving your standing in the community as a thought leader by proliferating the content you create. Content is king in this game.

CHAPTER 3:
SEARCH ENGINE ELEVATION STRATEGIES

If you are reading this book because you want to get involved in social media and don't know where to start, skip this section and head to Chapter 7, Social Media Suites. If you are already involved in using social media and want to find ways to develop, *optimize,* or, as I prefer to say, *elevate* your page in search engines, or are curious as to why you are paying someone to handle your *search engine elevation,* use the following information as a reference. This way, you can make sure your developer is doing the right things to optimize your website, and you will better understand what you are paying for.

On-Page Optimization

We will speak of keywords and key phrases often. Keywords and key phrases describe the word or phrase that someone enters into a search engine to find the information he or she needs. If, for instance you are an allergist interested in getting your website to rank higher in a search engine search, you first need to identify which keywords and/or key phrases drive the most traffic for allergy-related searches. The answer is going to be words or phrases that are relevant to the allergy conversation happening online, and that the words or phrases relevant to the allergy conversation are being incorporated into the *content* being published on websites, blogs, in forums, and on discussion boards, etc. If your content as an allergist contains many of the keywords and phrases people interested in allergy information are seeking, they are more likely to find you and, if the content reads well, will keep their eyes on your content longer. The search engines use software to figure out which content for allergy information is the most relevant and gets the most visitors, and this works in your favor when it comes to the ranking of your website on sites like Google, etc. If people stop to read your content but leave quickly because they are bored, or the content isn't well written or presented, the search engines can demote you. If your competitor has patient-education videos, patient encounter forms, or other tools embedded on their homepage and you don't, they might rank higher in a search engine, so people will likely find them first. Having good content, video, and forms, as well as other tools on your website or blog means you have good *on-page optimization*. On-page optimization occurs when you make changes in the computer code, or add content, tools, or utilities to your website or blog that search engines will likely give you search "credit" for.

Using On-Page Optimization to Improve Your Search Engine Position

Search engine companies are in business to make a profit, and it is a competitive business. Google, Bing, Yahoo!, Ask.com, and many smaller providers make a profit by drawing the most viewers to their search tools. The better search results they provide, the more people they draw; the more people they draw, the greater opportunity to earn revenue through advertising, including sponsored, banner, and pay-per-click advertising. The more useless information a search engine pulls up when someone searches a specific term, the less likely that person will return to that particular search engine to find information in the future, and the less likely that particular company will be able to compete for search space.

Google is the most successful search engine at this writing because they have a constantly evolving computer program, or algorithm, that enables them to offer the best search for their users. Google's algorithm responds to a user's request for information on a particular keyword by finding the most relevant information on the web for that particular keyword while suppressing information that appears to be irrelevant to the keyword search. Google also eliminates information from searches that appears to be attempting to play against the "rules" Google sets for placing higher in search results for the particular keyword we are talking about. This process of "pruning the information tree" results in a more specific search. Google doesn't publicize exactly how their algorithm works, but it does talk to the media and provide clues as to what one can do to increase the likelihood that one will rise to the top of the search results for the keywords one wants to be found under. If you are a florist, and you produce content on the Internet, and within that content use phrases that patients seeking floral services search for, you have taken the first step toward appearing relevant to a search engine, and you get relevance points for that. The more content you produce relevant to your subject matter that incorporates the keywords your potential customers are searching for, the more points you get; the more points, the higher you float in search results.

Another way of getting points is when other people recognize tour keyword-relevant content and "link" to it. If a company that manufacturers vases reads the florist's website or blog and sees their products mentioned, they might link their website or blog to the phrase in the florist's website or blog. This link is an acknowledgement that you produce keyword-relevant content, and the link recipient, the florist in this instance, gets a point for that. More links equals more points; Google's algorithm recognizes this and floats the content the florist created even higher in the search engine results for that keyword, adding relevance to the florist's efforts. The more relevant you appear online, the higher you place in the search engine hierarchy when you are categorized by the Google algorithm. In this way, the content you disseminate determines your relevance within a search for any particular keyword.

You can disseminate content by posting on your website, but the most popular way to disseminate content nowadays is with a blog. Blogging is easy, free, and has the ability to be regularly updated without having to pay a webmaster or developer to post the content you create. Getting your blog post out there is not enough—you need to proliferate it, and the tools that exist to help you proliferate content are—you got it—social-media suites.

On the social Internet, the "crowd" of people using social media are judges. They find your content and determine whether the information has value to them. If they like what you have to say, they may forward your content through their network. The more people who like what you have to say, and the more people exposed to what you have to say, the greater likelihood of them referencing your work by linking to it, which results in more points for you! More points = better positioning in Google searches for the content you hope to be found under.

Obtaining the Top Positions on Google

There are three ways to advance toward the top of the page on Google. You can use Adwords, the pay-per-click service offered by Google, and your listing will appear down the right side of the page or across the top in a highlighted color. You may appear on the Google map by structuring your Google Places listing or just be being where you are geographically. And then below the map are what are referred to as the "organic" listings, which is where you really hope to stand out. While an Adwords campaign can be effective, people tend to trust listings they find at the top of the maps and the organic search results because they know they are not advertisements. It's a win-win if your efforts float you to the top of the organic search, as you don't have to spend advertising money for this exposure, and people who find it are likely to trust it more than they would if you paid for that position. If you have more relevance points for the same search term than the competitor down the road from you, you will likely come up ahead of them in the free (organic) Google listings when someone searches for that term in your geographical location. This is the best possible scenario for your business—to float to the top of the organic Google listings—and the effort to float to the top this way is what I call practicing searchial media.

Search engine algorithms use many factors to rank content, and some of the more important factors (in no particular order) include page views; volume of incoming links from related websites; technical precision of source code; quantity of content; ratio of functional hyperlinks to broken hyperlinks; viewer traffic; revisits by visitors; time spent within the website; click-throughs; redundancy; relevance; advertising revenue yield; geography; and language. There are other characteristics, and we will touch on many of these later throughout the book.

Getting Found

PageRank

Google states on their Google Technology webpage that "PageRank continues to provide the basis for all of our web search tools." That makes it rather important. PageRank is a term Google uses to quantify websites visitors deem important. It does this by assigning a numerical value to every URL (the address of web pages on the Internet) it finds. Every web page has some assigned points to begin with. Points are increased as other pages find that page and create links to it, as forms, video, and a number of other things are added, some of them within the *code,* or language, in which the website was created. There are more than two hundred ways to increase the amount of points your web effort is rewarded, and they fall into two main categories:

1. Relevance—Google sorts by relevance first, and then ranks by
2. Importance—based on keywords and content on page and their relation to the search term used.

The importance of a page will increase as its PageRank increases, but if the page isn't relevant to the user's query, it won't rank at all.

Why PageRank?

Think of a search engine as a library containing tens of billions of documents with no way to sort through them. New documents are added every day, and the pile of unsorted documents grows faster than the librarian can file them in a way that will enable easy reference later. You are a PhD student and your instructor wants you to find that research article on global warming, and she wants you to find it *now*. What's your strategy?

The problem sounds impossible yet is a close analogy to what happens on the Internet. The solution is a search engine categorization of these pages. Google does it with PageRank.

It's historically interesting to learn that the term PageRank, while applied to web pages isn't *named* for web pages but for one of the founders of Google, Larry Page. Google describes PageRank here: "PageRank relies on the uniquely democratic nature of the web by using its vast link structure as an indicator of an individual page's value. In essence, Google interprets a link from page A to page B as a vote, by page A for page B. But, Google looks at more than sheer volume of votes, or links a page receives; it also analyzes the page that casts the vote. Votes cast by pages that are themselves 'important' weigh more heavily and help make other pages more 'important.'"

Increasing your PageRank gets exponentially harder over time. It is much easier to increase your rank from a two or three than a six or seven, so you have to do more and work harder at getting your page to increase in relevance over time, and this might be achieved by getting many more inbound links than you had to before.

Dr. Alan Glazier / Website

"Toolbar" PageRank versus "Real" PageRank

Google uses a small graphic bar known as a "toolbar" that is used to demonstrate what a particular piece of contents PageRank is. The PageRank toolbar can be found within the Google toolbar and the Google directory. If you hold your mouse over the PageRank graphic in the Google toolbar, it displays Goggle's PageRank of the page open in the browser. This represents Goggle's view of the importance of that page relative to other pages on similar subject matter. The view in Google directory doesn't display the numeric pager rank, only the graphic bar that demonstrates PageRank importance. While Google has not openly discussed all the details that go into rating via PageRank, it is a commonly held view that Google uses a rating known as "real" or "raw" PageRank to actually classify web pages, and not exactly what they display in the toolbar.

Real PageRank uses a scale of 1–10 (or 11, depending on your position in the search), and the scale is logarithmic, so as you move up the toolbar scale, the range of real PageRank that each toolbar value contains increases.

How Google Structures Your Rank

We already discussed the points you need to earn to gain better organic search position. These points are compiled and used to calculate your PageRank. Google software, known as "bots" (a word used to indicate a software "mini-robot") is sent out to perform a process known as "web crawling." The bot crawls different web addresses (URLs) and stores a copy of the pages it crawls in the Google "storeserver." The storeserver compresses and stores the pages in a repository. In the repository, there is indexing software, which parses the pages in the repository and stores the link in a file known as an "anchor file." (An "anchor" is the visible, clickable text in the hyperlinks, or blue highlighted text you see when you are on a web page.) Out of this file, the link database is generated that is used to calculate the real PageRank of your web page.

Here are a few things you should know about the toolbar PageRank and real PageRank values:

- If a page is created, and it immediately acquires a large number of inbound links, the toolbar PageRank value isn't going to reflect that until the next time Google updates their toolbar PageRank numbers

by sending out their bots, but the page will still receive the ranking benefits from those links, whether the toolbar shows it yet or not.
- **PageRank Still Doesn't Measure Relevance**—PageRank as a stand-alone metric is more of a measure of importance than relevance. Therefore, the keywords on a given web page do not affect its PageRank value. PageRank is increased by getting more links to the page, not by adding more relevant keywords. A relevant keyword without a link does nothing to add to your PageRank.

The PageRank system literally ranks every page on the web, placing them in order of most important to least important for specific keyword searches. Linking more than once from one page to another won't help increase the destination pages relevance more than the first link already did.

Getting People to Land on your Page Website

When PageRank occurs to your web page, it is distributed over the links going out and accumulated from the links coming in. By distributing PageRank over outbound links, you are not diluting points, or PageRank. If many pages are linked to page one, and it is important because of that, when it links out to other pages, it isn't giving up any of its importance, just spreading the information further!

Optimizing PageRank for your Website

If your website contains twenty pages, the PageRank flowing through cannot exceed a value of twenty. There are many factors that can hurt your PageRank and cause it to fall short of the optimal number. For instance, a PageRank that falls short might mean your website has poor *linking structure*. Dead-end pages are pages that have no outbound links—they are like a dead end on a street. By linking to dead-end pages, you can actually be penalized and have points taken away from your PageRank.

Your home page has tremendous value when it comes to linking relevance. Remember, your homepage is linked to all the other pages on your website, and it is likely you have other important pages on your website too, so linking your pages to one another is a good way to improve on your link structure and to build relevance.

Google wants to rank pages higher that are important and valuable to their users. If you are searching for something, Google wants you to find it so you go back and use Google the next time, and they want you to find it effortlessly, ideally in fewer than two clicks, so their definition of a highly relevant page is one that has unique content that is most-relevant to a keyword. Unimportant pages include pages with little useful content, such as "contact" and "about" pages.

Methods of Improving PageRank

One way to improve your PageRank is to have more pages. More pages = more content, and if there's one message I want you to take home from this book, it is that *content is king!* Of course the pages you create should have good content relative to the service or product you are promoting. Pages created that are outside the subject matter of the bulk of your site may actually cause you to lose relevance points, ultimately ranking you lower.

Subtracting Content

Site-level PageRank refers to the cumulative PageRank of your website or blog (the PageRank of each page added together). PageRank is also awarded to each individual page, or URL on your website. It is important to know that Google ranks pages, not entire websites, so it might be in your best interest to sacrifice site-level PageRank in order to increase a single page's PageRank, especially an important page.

The Google Webmaster Guidelines include the following recommendation when creating a hierarchy for a website:

Create a site with a clear hierarchy and text links. Every page should be reachable from at least one static text link.

Site Architecture Definitions

The following are important definitions for review:

Landing Page—A page that you are trying to get ranked for specific keywords. We want it to rank high in searches, so we want it to have a high PageRank.

Supporting Page—A page that is optimized for certain keywords, but not the page you are trying the hardest to get ranked for those keywords.

Global Links Navigation—These are the links that appear on the bottom, top, left, or right of every page of your website. It is useful for the text that you want to have appear throughout your website, such as your office phone number, and when you change it, the change automatically happens to all pages at the same time. Changing a global link is the fastest way to make changes on the site that affect its linking structure. We will discuss another way of obtaining relevance points, which is by frequently updating your pages. The bots Google sends out to categorize the Internet based on PageRank notice when something changes on your website and look favorably upon that; it indicates that your site contains current information, as opposed to it being a static page with old and unchanging information. By updating your Global Links and through other methods, you can continue to keep the bot gods happy.

Secondary Links Navigation—Similar to Global Links but may not appear on every page, just many pages. May also be important to quickly make a change on a website and affect linking structure immediately.

Depth of Content on a Website

To review, a page gains PageRank from its inbound links and distributes PageRank evenly across its outbound links. When a page receives a point from an inbound link, pages deeper and deeper in the website receive fewer than one point, so that page five clicks into the website will receive a fraction of the value of the inbound link that the first page the inbound link appears on will receive. This demonstrates why the important pages in your website should surface on the Internet first, and when creating your web architecture, be sure your web developer is made aware of the important pages. Remember, they know web development, but they don't know about your area of expertise and which pages might be important contextually to your visitors. I have a link to the homepage of my website on every blog I post. When people link to my website homepage, it gets the majority of the PageRank benefit, but there is also spill-over in terms of relevance boost to my other pages that are not dead ends.

Natural Versus Unnatural Linking

The language that a web browser understands is written in software code, and it is possible, yet undesirable, to write code that the browser understands but the search engine doesn't. It's not good to spend time proliferating content in a programming language that the search engine can't read and won't get you very far. Search engines have a difficult time interpreting code written in JavaScript (a standard functional programming language), while web browsers can interpret JavaScript, read the code, and create links from it. The search engines won't recognize it. If you have a link that is written in JavaScript instead of HTML (or HyperText Markup Language, the predominant "markup language," which uses a set of "markup tags" to describe web pages), the page would be prevented from distributing PageRank deeper in your website from that link, but the link would still work. This is a dead link, and it is important to locate your dead links, have them recoded, and make them live. Dead links are recognized by search bots, and your relevance score is penalized because of them. When you meet with your developer, make sure he or she understands this; it is a good tool to help guide your developers and monitor his or her knowledge and effectiveness. If you ask a developer about dead links, and he looks at you like you're from Mars, you probably want to take your check and run away as fast as you can. Most developers are ethical, but you are bound to run into those who know they work magic you can't understand. I want to give you tools so you can manage your developers, enabling you to end up with a better product for dollars spent on development.

Controlling PageRank with Software Code in Links

The links you or your developer create and place within your content have an effect on your PageRank. There are tools that can be put in place within these links to enable your important pages will more likely be recognized by the bots, and there are tools that enable you to post a link on a page you don't want to have recognized by the bots.

Here's why you would or wouldn't want that to happen.

"nofollow"

"Nofollow" is software code written in HTML language that instructs search engines that a link should not influence the link target's ranking in a search engine's index. In other words, it's a "stop" code and keeps the bots from discovering the link. This makes links essentially invisible to the bots that Google crawls with and doesn't take away or add value to your site's relevance score. Your developer would use this code, for example, to link to a dead-end page. By coding this way, the link to the dead-end page could exist, and you would not have relevance points subtracted for linking to a dead-end page, as the code would tell the bots, "I know I did this, and I'm telling you to bypass me as irrelevant." The purpose for the existence of nofollow is to reduce the likelihood of certain search engine "spam" from occurring, thereby improving the quality of search engine results and preventing spam from getting indexed. The fewer "spammy" links, the better the web experience for all of us, especially for the search engines. "Nofollow" is specific to Google and, as such, is a rule that might not apply for particular content when searched on competing search engines.

JavaScript

If you have pages on your website that are coded in JavaScript, a particular computer language, it may be hurting you. Web developers believe that while JavaScript was not recognized code in the past, that might be changing, and that Googlebot can actually read and interpret some JavaScript and is getting better at it. This is important for sites that are unknowingly preventing Googlebot from indexing their pages because they are peppered with JavaScript, but not good news for those sites that have used JavaScript to keep Google from indexing pages on its website.

Controlling PageRank With Software Code on Pages

If you have ever accidentally clicked on a link on the Internet and found a bunch of code, it probably looked something like this:

<a> TARGET "_cancer"/<BODY on Focus>< "can be a carcinogen" ></a>

For anyone who codes, this particular line of code makes no sense, but code that uses symbols like this is what's behind the text and pictures you see on web pages. Codes include "tags," code phrases that define the geographic regions of your website, where pictures are placed, where text is placed, and what colors appear on the page. Code is written with a "start" tag and an "end" tag, and the content is placed in between. Tags are used to specify what will take place on the web page when viewed and can be used to affect PageRank as well. Coding affects your web page indexing and crawling.

Indexing

Indexing is the process in which Google interprets the content of your web page and catalogs it for sorting in their index, processing it for search

35

results. When Google indexes a web page, it applies a value, called the value points earlier, to the document's relevancy for a particular keyword. When you go to a new search engine and "add" your site, you get indexed (hopefully) within that engine. Monitoring the indexing is important and should be done periodically.

Crawling

Google sends code through your web page, identifies the links, and uses the information to compare your page to others, find new pages, and ultimately calculate PageRank.

Robots Meta Tag

This is a tag is placed in the <head> (top) section of the code of a web page. It tells search engines whether or not they can crawl, index, or cache the content of that page. It is analogous to the nofollow we spoke of in the link section but affects the entire page instead of just one link. Robots meta tag is a "red light" or "green light" that allows the bots to proceed through your site.

"noindex" Code

If you apply the "noindex" code within your robots meta tag, you are telling Google not to consider the page relevant to any query and to not list it in any search results. This will affect your PageRank.

robots.txt

This is a tag you can add to your header code that does not prevent a page from showing up in Google's search results but blocks certain *users* from accessing certain files or directories on your site. It is a privacy control, and one that can be used to keep competitors from viewing certain aspects of your web presence. Applying this tag does not affect PageRank, unlike applying the noindex tag.

LinkDiagnosis (based on the Yahoo! linkdomain report) provides you with extensive linking reports for any domain. Your competitors have a method of gaining links from sources, and having access to the knowledge of "who links to them" and "who they link to" can assist your efforts in finding links. Understanding where their links come from can be of strategic importance.

For example, if your competitor has a link from a high level source such as a university and you don't, that link is providing value for their content in search. You might want to attempt to get a similar link from the source to level the competitive playing field for your similar content. The benefit of LinkDiagnosis is to help you learn who and what your competition is linking to . By understanding the strategies of a competitor, you can develop a more comprehensive and effective link campaign

Keywords, Tags, and Links

Keywords

Keywords are the words that best describe content you or someone else searches for. "NFL" would be a keyword that would bring up information on football—most likely professional football information—while "quarterback" will bring up information mostly about football in general. I provide these two examples to make the point that knowing which keywords are most important to have people find you in your particular specialty is key to your ability to be found in search engines. The best keywords or keyword phrases are not always the obvious ones. In my specialty, I thought that "eye exam" or "eye doctor" would be the most searched-for word or phrase, only to find out that the word "optometrist" was far and away the most searched term (at least last year) for people who were seeking eye care or information on eyes and vision. There might be hundreds of different keywords or keyword phrases people use to search for you.

Not sure which keywords to use? There are many tools to help you determine which keywords people are searching for your care use. Google has a keyword tool at Google.com/sktool, and a popular "finder" is keywordfinder.org. Google also has a keyword finder to help you find the best keywords to target your Adwords campaign (Adwords.Google.com/select/keywordtoolexternal), and there is a free keyword tool at wordpot.com. Beware—there are many paid services that offer to help you find the right keywords for marketing or social media efforts. You don't need them; as far as keyword searches go, free tools will tell you all you need to know.

Finding the Right Keywords

Keyword Tools

Typing in a keyword phrase in a search engine and clicking "search" brings up all the web content relevant to the keyword in order of relevance. Because of this, understanding which keywords to use to drive traffic to your website, to incorporate into your linking structure, and to use as tags is the most important part of your optimization efforts. Keywords need to reflect the terms people use most commonly to search for the service or product you offer. Having the right keywords brings pre-qualified people who are looking for the services you offer or products you sell directly to your site. This is free traffic. Having the wrong keywords is a recipe for failure that you cannot afford. If you don't use the words your patients use, your content won't be found. Don't think your knowledge of your industry is enough to choose the right keywords. You think differently than the layman who is searching for your products or services—your knowledge of your field isn't enough to choose the keywords that are best suited to draw from the general Internet searching population. You need help.

Here are some great tools to help you:

Keywordspy.com

Keywordspy.com allows you to discover the keywords your competitors are using in their SEO (search engine optimization) and Adwords campaigns. Data is updated daily, so you can be assured you have up-to-date stats. You can also find out how much your competitors are spending on their Adwords campaign, which can be incredibly useful. They also offer a tool to help determine what some of the best keywords may be for your campaign. They have a tool that allows you to do real-time tracking of your keywords in the major search engines. Some of the other metrics you can compare include competitors' advertising copy across the big three search engines; Google, Bing and Yahoo, title, meta keywords, and landing URLs. They provide you the ability to monitor pay-per-click bidding on the keywords you use and mine their total keywords and ads. It is an excellent tool.

Google's Keyword Tool

Google's free search-based keyword tool provides keyword ideas based on actual Google search queries, matched to specific pages of your website and can provide ideas that are supplementary to your existing Adword keywords being used. There's also a great video on the homepage that discusses keyword choice and optimization. Adwords also has its own keyword tool as well.

The Google Trends tool offers:

1. Search term trend research
2. Website traffic analysis

Google Trends combines information from Google search, Google analytics, and third-party market research to help you find trends and analyze your website traffic.

The Google Trends keyword service shows the amount of search on the Internet for any particular search term across the world and in different languages. It enables you to use the following inputs (also known as operators) to find the following specific information:

- [keyword1, keyword2] to *compare* two terms
- [keyword1 | keyword2] to check for *either* of the terms
- [keyword1 - keyword2] to *exclude* one of the terms
- [(key phrase) | keyword] to compare several search terms, one or more of which is/are *multi-word term*(s)
- [key phrase] to search for an *exact match* (suitable only for two or more word combinations)

SEO Book Keyword Suggestion Tool (tools.seobook.com/keyword-tools/seobook/)

This site tries to sell you a lot of stuff, but registering is free, you can try different keywords for free, and it spits out tons of data, such as daily estimates of Google, Yahoo!, and Bing (all major search engines) "'hits" on the keyword, and it subsequently lists related keywords and their popularity.

Wordtracker

This is a free keyword-lookup tool. It also provides a list that shows exactly how a keyword appears in other searches. Wordtracker has an easy to use, effective, and intuitive search module and is a good place to start finding your keywords.

Ad-Center Add-In for Excel 2007

This product by Microsoft is a keyword research and optimization tool based on Microsoft Excel, which allows you to build keyword lists, plan keyword strategies, and forecast keyword impressions when you use Adwords or other pay-per-click campaigns.

Term Extractor

Many companies offer tools called "term extractors"—a "library" for taking natural text and extracting a set of terms from it that make sense, without adding additional context. It helps you discover the phrases that people use certain keywords within, and perform phrase-based keyword optimization. It applies certain weights to HTML elements and other on-page factors to determine what it thinks is a targeted term.

Keyword Discovery

This compiles keyword search statistics from more than 180 search engines worldwide, to provide a powerful keyword research tool.

Other Keyword Research Tools

Wordze.com and NicheBot.com provide keyword research that gives you access to more than 295 billion keywords in forty-one languages spanning 243 countries using multiple keyword sources.

Google Trends

Google Trends allows you to compare the world's interest in the topics you are curious about. It allows you to enter up to five topics and see how often they've been searched on Google over time.

Soovle.com

A search engine described as being for people who don't know what they're looking for. As soon as you start typing a search term, Soovle begins to offer suggestions for related terms that may apply to your search. It covers seven web services, including Google, Yahoo!, Wikipedia, Answers.com, Ask.com, YouTube, and Amazon. The interface is very much like a Google homepage.

Where to Place Important Keywords to Maximize Search Relevance

Tagging

"Tagging" is the term given to phrases inserted within the computer code of your web page that give clues to what your page is about. Tags can be applied outside of the computer code (which will be discussed when we cover blogging and social bookmarking). The computer code tags are an important part of the source code of your web page, and just like well-tagged pages can produce excellent search results, poorly tagged pages can kill your efforts. Tagging is done using words that are "keys" to people finding you and your efforts, be they service or retail, so we refer to these words as keywords. Your tags need to reflect the keywords that most people use to find the services or products you are offering. If your website code is not tagged with the right keywords, people searching for what you are selling won't find you easily.

Another subject we will cover are "inbound links." Inbound links are connections between your content and other content on the Internet and are viewed by search engines as incredibly important when determining how important your website is in relation to others for particular keyword searches. The more important you are, the higher you appear in the rankings when someone searches for—to use me as an example—an "optician" in "Pittsburgh," if people link to you using these keywords. Links are strengthened by using tags and keywords.

Title Tags

One of the four most important factors in raising search engine ranking is using tools called "Title Tags" (TTs) properly. The other three factors are (1) getting inbound links (2) producing content with specific key words and key phrases and (3) Social Influence scores. They are just as important as the text you post and the links from other websites that point to your pages (inbound links), if not more so. Altering the TTs within the code that creates the pages of your website can have an immediate impact on your search ranking. *The words placed within the TTs on your pages are what appear in the clickable link on a search engine page.* If you search

41

for me using the phrase "optometrist in Rockville, Maryland," the first link that comes up in the organic (free) listings in Google is "optometrist in Rockville-Shady Grove Eye and Vision Care" (at least at the time of this writing). If the phrases you use for your title tags don't reflect the search terms people in your geographical vicinity most use to search for the product or services you offer, they should be changed, and you should see an increase in visitors in a very short time.

Title tagging is something that is often overlooked by webmasters and others hired to optimize their website to be found in search. It is highly recommended that, when hiring a webmaster, you oversee his or her efforts in terms of what title tags he or she places. Remember, this person is not in your industry and probably needs some hand holding to know exactly what terms are going to drive the type of clients you are looking for to your business.

Every page on your website should have a unique title tag that utilizes two to three key phrases, separated by hyphens, commas, or pipes (pipes are the | symbol used in HTML language to denote a separation of two or more words, phrases, or characters).

Should you Put Your Company Name in the Title Tag?

It is okay to put your company name in the title, and you can place it at the beginning of the tag. If your company is a well-known branded company that people search for by name, it is even more important to use the company name. If you are not particularly branded in your market, title tagging with your brand is a good way to build the brand.

It is important that your title tag contain more than just your company name—you also need phrases that are descriptive of the services or products you offer. There will be people searching for you by your name or brand, and those who have never heard of you will find you because your TT reflects the services or products you are offering and they are searching for.

I believe that, as you attempt to optimize yourself through search, the services or products you offer are more important for your title tags, as your efforts are really more about driving in new business than marketing to existing customers or those who already know about you and how to find you.

Searchial Marketing

Keyword Phrases for Title Tags

If your company's name is Jane Doe, attorney, and you were located in Gettysburg, Pennsylvania, you would want your company's website to appear in search engine results for searches for "Gettysburg attorney" or "lawyer Gettysburg, Pennsylvania." The more specific your search phrase, the better your search results will be. If you use the TTs "lawyer in Gettysburg," you are more likely to see greater results in attracting new business than if you use the phrase "lawyer in Pennsylvania." More Pennsylvanians will find you with the latter, but your TTs are less targeted—and it's probable that people from Philadelphia would click, but how many would end up in your office?

Your tag might appear like this:

Jane Doe—lawyer in Gettysburg

Or you might try

Jane Doe—lawyer in Gettysburg Pennsylvania

But you have enough space to include both of these in your tag

Jane Doe—lawyer—civil law in Gettysburg Pennsylvania

The idea is to write compelling titles as opposed to boring factual titles when possible. You can pretty up our last example like this:

Jane Doe—lawyer (civil law) serving Gettysburg Pennsylvania

Crafting title tags is important, but the best way is not determined by how you think they look. You should test them to see which tags bring the most traffic to you.

Key Phrases

In order for the bots to recognize your pages, every page on your website or blog should have a page name identified by a key phrase relevant to the subject matter of that particular page. For example, a page on my website youreyesite.com that talks about myopia control should have the page name: myopia-control.asp.

43

The URL should include the page name like this: http://www.youreyesite.com/myopia-control.asp (assuming "myopia control" is the highest-ranked keyword phrase to describe the subject on that page). Use hyphens to separate words—Google ignores underscores!

To maximize relevance points by page, every page should have minimum two or three paragraphs' worth of keyword-heavy text. (Text embedded in images doesn't count toward this rule-of-thumb.)

Short-Tail Versus Long-Tail Key Phrases

Short-tail key phrases are broader key phrases that encompass a less specific definition of the content being searched, while long-tail key phrases are more narrow and specific in terms of content the user hopes to bring up. A short-tail key phrase might be something like "physician Phoenix," while a long-term key phrase might read "physician pediatrician Phoenix Arizona suburbs."

If someone searches for "physician Phoenix," he will come up against many businesses competing for the same views. There is also a much wider distribution of people searching for a two-word key phrase, such as physician Phoenix, than a three-word key phrase, such as physician pediatrician Phoenix, and even more so than for a four-word key phrase, such as physician pediatrician Phoenix Arizona. When you compare how many people search for the two-word key phrase to how many search for the four-word key phrase, you find an inordinate amount of people searching for two-word key phrases in comparison.

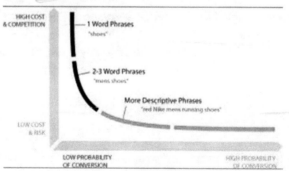

source: http://www.helloloscabos.com/seo/long-tail-keywords-vs-short-tail-keywords/)
credit to Elliance, Inc.

The advantage of using long-tail key phrases is that the people using them convert at a much higher percentage to clicks. Those searching for long-tail key phrases seem to know exactly what they are after, thus are searching the Internet with a more defined purposes, and businesses are more likely to convert their search into a click, meaning real business. This does not mean that one should focus exclusively on finding long-tailed search terms, but it does suggest that it is worth the time, effort, and imagination to devise several long-tailed key phrases that your customers might be most likely to use, and implement them as attributes when you tag and code your website or blog.

Use Your Visible Text Copy as Your Guide

It is recommended that your TTs reflect the copy on the page. For this reason you should analyze tags after the web page has been written and optimized. The phrases you use in your copy and how you tag them play an important role in optimization, so TTing based on your tagged text will give you another boost in the engines. You should use the most important phrases in your copy and tag each web page using these phrases.

Auto-Generated Title Tags

Sometimes blogging systems, such as Wordpress, will use the text from your page to automatically generate their own title tag for you. I recommend you hire a webmaster to analyze different title tags that may have been generated by these "content management systems" and perform a "work around," i.e. change the structure as much as possible to alter the tags as the generated title tags might hurt your efforts. If your webmaster says this can't be done or doesn't know how to do this, it's time to find a new webmaster.

How to Make the Best Title Tag Possible:

1. **Use Your Brand**—Use the name of your site or business at the beginning or end of each title tag. This helps your visitors know where they're going and remember where they've been. Remember, searchers will look further down the organic search rankings to find a trusted brand.
2. **Use No more than Sixty-Five Characters**—While most engines have supported title tags up to sixty-five characters, Google now supports seventy characters, and this might change for other search engines or

increase with time. The reason to limit them is search engines won't look at tags longer than 65 or 70 characters

3. **Use most Popular Keyword Phrases**—Use keyword research to determine the most popular keyword phrases and *use them!* They work.
4. **Use Long Descriptors with Pertinent Phrases**—There is an art and a skill to writing title tags with good descriptors. Use as many as are relevant to the page and the search being done. A phrase like Opticians/Designer Eyewear in Pittsburgh Pennsylvania is a better descriptor than Opticians/Designer Eyewear. You might have a web page where the title tag is more relevant to designer eyewear, and a landing page for opticians Pittsburgh, so you don't want to create title tags that split the traffic between two of your pages as it will lessen the effectiveness of the tags on both pages, decreasing relevance within searches for both phrases.
5. **Separate Terms with "Dividers"**—Commonly used dividers include the | symbol (a.k.a. the pipe bar), the arrow (>), or hyphen. I have read the suggestion that the arrow or hyphen is used inside a title tag, as with a title, so it would look like this: *Eyewear | Articles > Keyword Research - A Beginner's Guide.*
6. **Benchmark using Click-Through and Conversion Rates**—Use your analytics to determine "click through rates" (CTRs) as a metric. You shouldn't depend entirely on CTRs as the sole measure of success for your campaign; you need to take into account conversion rates (the ratio of visitors who convert website visits into action. For example, it's a "click through" when someone ends up on Amazon.com from another site, and it's counted as a "conversion" when they buy something). It's okay to have a low click-through rate if your conversion rate is higher.
7. **Target Searcher Intent**—Your titles should be specific to the intent of those searching for you. If you are marketing your eye care services, you would want your tag to read Optometrist/Determine Your Risk for Developing Macular Degeneration, whereas if your intent is retail-oriented you would want the title tag to read Optometrist/Fashion Frames in Pittsburgh.
8. **Maintain Consistency**—When you have found a system that works for you, repeat it with consistency. Over time you will build a searchable "brand" and develop a following.

Search Attributes

The code behind your website, usually but not always written in the HTML computer language, also called "HTML element" has attributes that provide additional information about the element in which they are placed. Attributes are always specified in the start tag.

Meta Tag Attributes

"Meta tags" contain information about the web page that search engines use to help organize and categorize the millions of web pages out there. Each page's unique "meta" description tells the search engine what your page is about. They are usually three to five relevant, targeted key phrases placed within the computer codes that are programmed for your web page but are not visible to users viewing the page, and are used to describe your site in search results. If you do not put a meta tag as a page descriptor, what appears in the listing would likely be whatever first appears on your page, usually the alt text of some graphic or banner, or perhaps your top menu.

In the early days of the Internet, meta data was the major metric used to assist in search engine elevation. Having good meta data frequently meant having a higher search engine ranking, thus more website traffic to those who used meta tags correctly. In the more modern Internet, search engine optimization is much less dependent on meta tags. These elements have decreased in value, as they were unscrupulously used by website content developers to get around better accepted ways to get to the top of search engine rankings.

When hiring "experts" to increase your search ranking, ensure your developers won't be using unscrupulous methods to get you there. You want to make sure that your developers don't do something to your code that associates you with keyword spamming; if you don't, your efforts might be in vain; for every relevance point you pick up for doing something right, you may lose points if your content is linked to broken, dead, or spammy links.

Today's search engine bots like to see your web page have other characteristics to be ranked higher, such as incoming links, good content, technically appropriate source code, links that click through to somewhere (functional links, as opposed to broken links), and many other metrics that will be discussed throughout the book.

The Keywords Attribute (AKA Meta Keywords Tag)

The keywords attribute, at one time, was one of the most commonly used meta elements. In the late '90s, search engine providers realized that information stored in meta elements, especially the keywords attribute, was often unreliable and misleading, and at worst, used to draw users into spam sites. (Unscrupulous webmasters could easily place false keywords into their meta elements in order to draw people to their site.)

Nowadays, however, major search engines like Google and Alta Vista ignore this tag, and the keywords tag no longer has the significance it used to have in the early days of the web.

Yahoo! is one of the leading search engines that claim to support the keywords meta tag in ranking your page. Google claims they don't use it at all.

Important: *Don't use meta tags or other elements to include misspellings of terms you hope to land visitors from.* For instance, if you are attempting to create content to compete in the eye care space, don't tag things with different spellings of "ophthalmologist" (such as opthomologist or opthamologist). Search bots can recognize these and actually penalize your site for using misspellings, even if you have good intentions. You should also not repeat a particular keyword too many times in your tag, as it may be recognized as spam and get booted out of the searches you hope to be found in.

The Description Attribute (AKA Meta Description Tag)

The description attribute is a concise summary of the content available on the web page. The description attribute is supported by most major search engines. The description attribute allows those authoring a web page to provide their own description of their page, as the search engine might automatically produce its own description of the web page, which may be less accurate. As the description is often displayed on search engine results, it can affect click-through rates. Unfortunately, some of the free blogging platforms like Wordpress won't allow you to enter a meta description tag, which would be to your benefit.

The Language Attribute

The language attribute is a clue for search engines to understand which language the website is written in (not coding language, but languages like English, Spanish, Hindi, etc.). It helps tell search engines which language a page is written in.

The Bots Attribute (AKA Meta Bots Tag)

This attribute allows or doesn't allow search engine bots to index (list) a page within the search engine. Using this attribute, you can control whether your site is or is not indexed (noindex), or crawled ("nofollow"; see discussion of nofollow in Chapter 6). There are other attributes, such as "noarchive," which tells a search engine not to archive a copy of the page, and "nosnippet," which asks that the search engine not include a snippet from the page along with the page's listing in search results.

Meta tags are not the best option to prevent search engines from indexing content of your website, because a more reliable and efficient method is the use of the Robots.txt file (robots' exclusion standard). It is important if you want certain search engines to crawl through your website using the links on the page you submit.

Building Relevance through Linking

Your content will be linked to, and you will "link out" to content. Creating links is a great way to promote your efforts and often results in people linking back to you. As you know by now, the more people who link to you, the more relevant you appear to the bots. You will be creating links from your content to other relevant content most likely within your blog. Creating links on the website usually requires a developer, but free blog software enables you to do this in a platform that is both easy and intuitive. Linking out to other sources can reflect positively on your search engine rankings and help to build trust between you, those who wish to link to you, and people who are viewing your content.

A survey of major newspapers on the web (http://www.seoco.co.uk/blog/how-good-is-the-mainstream-media-at-linking-out/) found that those that link out tend to outperform those that don't on many performance metrics. Think of sites like Digg, Reddit, Yelp, Twitter, and Delicious—all of which link off their own sites as part of their core business and still get visitors coming back again and again.

Benefits of linking out include:

- **Creating Traffic**
  People who track website traffic will notice links, even if they aren't clicked on too often. Noticing links brings attention and attention brings visitors.
  It Makes your Site a more Valuable Resource
  Links to sites within your field with slightly different expertise, or sites that contain much more information than yours on similar subjects increase the value of your site from the perspective of your site being used as a resource by others. The more intellectual value, the more value your website is perceived as having to users, creating credibility that translates to higher search ranking.
  Search Engines Reward the Behavior through Better Algorithmic Positioning
  Just like the people you keep company with can reflect the type of person you are, the links you send are used to signal the level of quality of your website. Good sites typically link to good resources, and spammy sites tend to link to spammy resources. Engines pick up on this, and websites that are viewed as trusted sources of information are given a greater "pop" than those that aren't.
- **"Build Links and They May Come"**
  When you link out, it sends the message that your are participating in the linking environment going on in the web. It signals that you are willing to converse and help out and be helped out. When you link out, you will get some interest in those you link to linking back to you, which adds tremendous value to your website in terms of web search relevance. In participating, you provide the "linkerati" (people who are busy linking around the web, such as bloggers, social media sophisticates, online journalists, website builders, and forum participants) reason to link back to you.\

Cross Linking

Cross linking refers to the creation of links that link to one another within the same website. It is a way of leveraging your own website to increase your link profile

Cross linking is important if you intend to share PageRank with all of your other web pages. The more your pages share PageRank, the better optimized they will be through a Google search. If you pursue an aggressive cross-linking strategy, you can expect to watch your PageRank increase rapidly, and you will also see your pages show up more often in search engine results pages (SERPs).

Search engines like keywords in the hyperlink (anchor) text better than those in the rest of your content, so links containing relevant keywords to your page from outside domains help even more.

It is in your best interest to link as many pages within your website to the pages you are optimizing for search engine position. Following that, be sure they all get indexed (see Chapter 6 for more information on indexing). When you crosslink within the content of your own site, search engines have an easier time crawling through your site, and this results in higher visibility, or "float," of more of your pages in search engines.

Google requires at least one link to your website from an outside website or domain for your page to be indexed, so be sure there is a link on each page of your website to and from another domain.

The cross links you create should occur within the content of your pages and not within the footers of your home page. Use keywords and key phrases to create the links, not headers, graphics, or pictures, as these links don't get searched by the bots.

Links must be made logically and follow the ideas within your website. Too many crosslinks can result in being penalized in the search engine. Avoid crosslinks that don't make sense, such as linking a word like "nosepiece" to a word in a piece you wrote on nutrition.

Dividing and Cross-Linking Content (Spidering)

This strategy for search engine elevation involves owning several domains and creating websites with keyword relevant content, splitting content between the domains, and cross linking the sites to one another. This can be helpful to your search elevation pursuits, but there are some rules: If you practice dividing and cross linking, be sure to host each site on a different hosting service. This way, each domain has its own IP (a unique, numerical label assigned to each device [e.g., computer, printer] participating in a computer network that uses the Internet Protocol for communication) address. Today's search engines are programmed to recognize what is referred to as "neighborhood" cross links, and cross linking between domains owned by the same person, and they look unfavorably upon that—they might penalize you in terms of your ranking within a search. Each domain should have completely unique content, as bots can easily pick up on repetitive content between websites. Be sure each website has "depth" (e.g., more than a few pages but not too many—five to ten is good). Search engines like websites with depth and will often ignore sites that are too shallow.

Try to create links from a page on a separate domain that include your main keywords in the title or header tag—this will improve your link popularity. The more relevant one page is to another, the more the particular crosslink increases in value in terms of benefiting your position within a search engine, thus the more relevant the pages become.

Cross Linking and Popularity Rating

Outbound links are links that point at someone else's content and are important in your efforts to obtain reciprocal links. People see you linking to them and may link back out of interest in your subject matter or out of courtesy. There is a downside to this strategy: each outbound link you create dilutes the "popularity rating" (PR) of the page you are linking from. Remember, the more outbound links on a page, the less percentage of its total PR will be transferred to your page. Obviously, the PR transferred in the end will ultimately depend on the PR of the page that links back to you, the number of outgoing links it has, and its relevancy.

Cross Linking and Anchor Text

Google searches for pages that share keywords with our page, not just in the header or title but in the content, specifically preferring keywords in the hyperlink, or anchor text, itself. The crosslink increases in importance when it contains keywords that people are using to search for businesses like yours.

What you want is back links to your pages that share your keywords in the title and the header as well as the anchor text. The search engine reads this to mean your page is relevant to the query. The content must also be relevant, containing phrases and content pertinent to the conversation within the content for best results (see PageRank section, Chapter 6). Always make sure to use important keywords within the anchor text, title, and header tags to maximize results.

Link Attributes

Link "Anchor Text" (AKA "Link Label" or "Link Title")

Anchor text is the visible, clickable text in a hyperlink. The words contained in the anchor text can determine the ranking that the page will receive by search engines. Anchor texts normally contain fewer than sixty characters. Anchor text provides descriptive or contextual information about the content of a link and its destination. Anchor text linking is useful to users and webmasters as anchor text is looked favorably upon by search engines in determining relevance.

Common anchor text looks like this:

<a href="http://en.youreyesite.com/eyes/Main_Page">Youreyesite</a>

In this example, my website youreyesite.com is anchored in the awkward and long description but is displayed neatly as Youreyesite.com in the search engine.

Anchor text is usually written to describe the page it points to (landing page), and as such, is given great importance by search engine algorithms due to its relevance for what is being searched. Google will allow you to view the most common anchor text linking to your site using Goggle's Webmaster Tools. There are some important rules to creating these outbound links, so pay close attention!

Link Anchor Text

Links to external sites should consist of anchor text that includes the keywords of the page being linked to. Whenever possible, you want to link to pages deeper in the site and avoid linking to home pages that have the same site name as link text. For instance, if you want to link to my practice's website youreyesite.com, you don't want the link to include only the phrase youreyesite.com (the link to my homepage), you want to link to something like my page about myopia: http://www.youreyesite.com/ myopia, or macular degeneration, http://www.youreyesite.com/

macular-degeneration. Linking deeper creates more relevance points. Also, if possible, include only one link per page to the targeted/linked page. If you have multiple links to the same target page, try to use the same anchor text. If that's impossible, the link higher up in the code should contain the more desired anchor text.

Image Alt Text

"Alt" text, short for "alternative text," is text that provides the same information as the text in an image on a website. It is typically brief and coded for the image. An image that is decorative typically doesn't need image alt text, but an image that conveys content, such as a graphic, would.

Body Copy

Body Copy refers to the content we find or produce in website publication. It is the "meat" of the stories and articles we produce and read. If you are an accountant and write an article on the latest tax laws, the body copy is the entire text of your article.

Your content shouldn't be too long. A good rule to follow is to not write a blog that is greater than two hundred words on a page. You should be sure to include plenty of searchable keywords spaced evenly throughout your content, but not so many that visitors get the feeling they are being marketed to. The content needs to be interesting enough to keep viewers on the page longer, thus on your site longer. For blogging, brevity is key—people don't have the greatest attention spans, and experimenting as a wordsmith by writing long documents might cause people to lose interest and click away from your site, increasing the "bounce rate" (see Chapter 16 for more information on bounce rate).

Keywords should be used in the beginning, middle, and end of the content to create a theme. Search engines pick this up easier when people search for associated content, and this will improve its ranking. The first sentence should include the most important keyword you expect will be searched. Don't forget to have the keyword(s) in the final paragraph as well. Some search bots start by searching the first and last paragraph of your text. Keywords should be as specific as possible to what you expect people to search for when they are seeking the information you publish.

It is likely the first few lines of your text will be used as descriptors by search engines, so make sure they accurately convey the message you want out there. Think of the first few lines of your text as your "pitch"—people are going to come, or not come, based on how interesting, believable, and pertinent your search engine pitch is.

<H> Tags (AKA Heading Tags)

Headings (or *head* elements) are typically titles that describe content that is to follow. Headings are placed with code that is written to increase the importance of your heading when searched by bots. The head element is code buried within your web page that contains information about the current document, such as its title, keywords that may be useful to search engines, and other data that is not considered document content. Every HTML document *must* have a title element in the head section. The title element should contain the most important keywords relevant to the searches you hope will lead to your landing (main) page.

Headings are important in technical or business writing as they alert people to the content they are about to encounter. Their usefulness in technical writing makes them an important subject when relating them to use for the professional. A poorly headed technical article can turn people off from reading your content, yet a good or interesting heading will frequently result in a customer delving deeper into your content or website. When dealing with medical or scientific subjects, you should pay close attention to how you head your content; the layman might be quick to turn away from too technical a heading.

Heading tags are used by search engines to identify words that are more important than the rest of the page text. Search engines use your headings to index the structure and content of your web pages. In search engine elevation, it is thought that the heading is highly weighted by the search engine when it sums up the topic of the page, so the heading is considered an important keyword.

Screen readers and magnifiers (for the visually impaired) rely on headings to navigate the page, particularly important for ophthalmology, optometry, or opticians. If you post information for the visually impaired, be sure your headings are descriptive and simple, or your target audience might never find them. (An aside: one of my patients, a sixteen-year-old high school student at the time, created a search engine that searches information on the web based on font size. The search box itself is larger than Goggle's standard search box. It is useful for the visually impaired as well as for people suffering from computer vision syndrome. This new search engine is now available for the public at http://www.good50.com.)

Blogs

When the search bots go out, they scour your site for certain things. One main thing they like to see is a blog associated with your website. Big searchial points for that, by the way. Blogs are tools that help increase relevance by allowing you to update content regularly without the hassle and expense of involving a web designer or webmaster. Content on the Internet, including within websites and blogs updated more frequently with subject-relevant information, are perceived as more relevant to a given search, and search ranking goes up. If your website does not have a blog, you are missing a major tool to help elevate your search position. Blogging can take as many as three months to show palpable results, although in fewer than two weeks, it's likely your content will get hit by a bot and be indexed in Google. You have to post regularly (three to four times per week) to start seeing results, and use social media to market and proliferate the link to your blog posts (a lot more about that later), but patience is rewarded, because in the end, the search engines index your blog and each of your posts. So it is like having a second website promoting your business or products, you compete better in the search engine, and a blog increases your reputation and relevance in the space on the Internet in which you conduct business. (See Chapter 6 for more on blogging.)

Website Forms

Websites generally have contact information posted conspicuously. This can be fashioned in any way, from a simple e-mail link to a detailed form the user needs to fill out. Many websites include contact information in the form of an address, e-mail address, telephone number, etc. There are benefits to simplicity, but in the modern Internet, or Web 2.0, a simple method of contact, such as phone, address, or e-mail, is not enough. Every modern website should have a contact form that not only allows people to contact them but collects data on the visitors so they can be marketed to and trends can be studied, so the habits and information provided by users can be mined, enabling action can be taken to stroke existing business.

Forms can help you better manage your business by providing clients a method of contacting you, whether it concerns customer-service issues or suggestions as to which direction they would like to see your business head in. They can provide feedback in many ways, and the astute manager will find value in feedback as another angle to understanding his or her business. Forms allow you to maintain a closer relationship with clients during times they are not patronizing your business. Without forms, your whole web effort is lacking an important means of contact between you and your clients. The ultimate goal of any website is to get more visitors and then more conversions from visitors to clients. A static page, where nothing happens, or people are compelled to go through too many steps to get in touch with your business is sure to cause an abandoning of your site by those you might have had the opportunity to push information to or glean further growth from based on their feedback.

Your contact form should be simple—you might turn visitors off with too complex a form, negating your efforts. Have a system in place to quickly respond to form queries—this is an important task to delegate. Test the forms to make sure they are working regularly, and test them to make sure they work in different kinds of browsers; for example, forms created in Firefox might not function in Bing. At our practice website, http://www.youreyesite.com, we have patient registration forms patients can fill out before they arrive, saving time when they get to the office. We also have appointment scheduling forms and HIPAA forms for their benefit. These forms help with our PageRank while providing a benefit to our practice and to the patient.

Bots love forms embedded within your website and reward you for having them. Forms add value to your website from a customer and a search engine perspective. Search engine bots can dramatically increase your search relevance points when they crawl and discover you have forms on your website.

Images on Your Website

Images on your website can be a double-edged sword. Good images make the website attractive and fun. For medical sites, they are likely to educate as well. Images brand, and provide maps and other descriptive elements—everyone loves a visual. The downside to images is that certain file sizes may slow the downloading of your website, causing visitors to leave early and resulting in a less-than-effective marketing tool.

Dos and Don'ts of Putting Images on Your Website

You don't want much of the time your visitors come to your website to be spent uploading major graphics, including static images. You can improve on your page upload time by reducing the number of components on the page, including photos, flash images, or pictures. There are several ways to do this:

1) Create an **image map**—Image maps convert multiple images into one image, reducing the total size of the files on the page.
2) **"Cascading Style Sheets" (CSS)**—Style sheets are programs that define how documents are presented on screens. Remember the days where you would upload a website and would see pieces of different images come up bit by bit? It was programmed to appear like that to make it look like things were uploading faster, when, in fact it just slowed things down. Each upload on those old images was a separate "request," and the more requests, the less efficient the upload. CSS sprites is programming used to combine background images into a single image, enabling faster uploading and ultimately frustrating people less; the less frustrated they are uploading, the more time they'll spend on your site.
3) **Inline images**—Embedding images in the actual page code can increase the size of your content page your visitors are trying to upload—which is bad. Combining inline images into stylesheets is a way to avoid increasing the size of your pages so they upload quicker—which is good. Inline images are not yet supported across all major browsers, so while they may open in, say, Firefox, they may not open in Internet Explorer or vice versa, so if you want to maximize your exposure across the web, this method is not yet ready for prime time.

Hypertext Transfer Protocol (HTTP) Requests

Interactions within the Internet occur when you have "clients," usually people or computers that are requesting a "service," and "service providers," usually computer servers, all taking place through a computer network. This is called "client-server computing." HTTP describes the message exchange pattern that occurs on the Internet, where someone acting as a requestor within a system sends a message out in the form of a command received by a "reply system" that processes the request and, hopefully, returns the message. Thus, the Internet conversation happens. This pattern is described as Internet "architecture," and HTTP is the most common type for standard client-server computing used today. For the purpose of our discussion, an HTTP request is simply a message in the form of a link placed within your content (HTTP://www.*something*.com).

A way to improve the performance of your content is to reduce the number of HTTP requests in your page. This is the best way to improve upload performance for first-time visitors. HTTP "cookies" are pieces of text that your web browser (e.g., Internet Explorer, Firefox, Chrome, et al.) stores after you have visited a website. Cookies act as memory—once you've been to a website, your computer may store a cookie of it. Having a cookie in your browser for a particular website helps the site to upload faster the next time you visit the site. The cookies are stored in a "cache." The cache stores cookie data so future requests for that data can be served faster. As described in Tenni Theurer's blog post *Browser Cache Usage—Exposed!* (http://www.yuiblog.com/blog/2007/01/04/performance-research-part-2), 40 to 60 percent of daily visitors to your site come in with an empty cache, meaning they don't yet have your contents' cookies, and uploading without existing cookies takes much longer than a return visitor to your site who already has cookies downloaded. Making your page faster for these first-time visitors is key to a better user experience.

To find out how fast your website loads, visit http://www.seo-shop.com/tools/tools/website-speed-test/. Visit these sites for information on how you can monitor your developer and make suggestions for speeding up the upload time on your front end. http://developer.yahoo.com/performance/rules.html and http://developer.yahoo.com/performance/rules.html

1. PhilYeh.com

CHAPTER 4:
OFF-PAGE OPTIMIZATION

Off-page optimization refers to ways you can improve your position in search engines by doing things outside the scope of your website or blog (e.g., by not changing the architecture or content of your site). Successful off-page SEO will result in your receiving significantly higher PageRank on search engines like Google, Bing, MSN, and Yahoo!.

Most successful off-page optimization happens because you take actions that result in people linking to your content from theirs. Bots *love* links to your website from sources that host content similar to yours. Bots don't like links to your website from spammy sites. There are many ways to achieve good links, many of which will be described later in this chapter.

Another off-page method of improving your ranking is to submit the content of your website page-by-page to directories, which we will discuss in greater detail later in this chapter. You can also participate in forums and social media websites; all help increase your visibility and profile within search engines.

A Do-It-Yourself Off-Page Optimization Primer

You can spend tens of thousands of dollars on "experts" to optimize your sites for search, but if you don't understand what they are doing, you may get less than you bargained for. If you do decide to hire developers for your off-page SEO efforts, which some recommend, I suggest you find someone who has more than two years' experience performing off-page SEO and has references from companies involved in Internet businesses. First, make sure *their* search engine listings are optimized. They should be able to get themselves to the top of searches, and if not, you should ask how they expect to get you there if they're not there themselves.

Here are some things you need to know about off-page optimization so you can practice it yourself or guide your web developer. In order to manage your optimization efforts off page, you should:

1) Perform keyword research and test best keywords for use
2) Incorporate keywords in link anchor text (see the Link Attributes section in Chapter 5)
3) Get high ranking websites to link to you that are one-way—directed at you, with no link directed back to them
4) Use different keywords in your links from the same site
5) Ensure your link building occurs gradually; search engines are aware when you generate too many links at once, and this can actually hurt more than help
6) Make sure keywords around your links have contextual relevance
7) Link deep into your website throughout all pages, even pages that seem of minimal importance
8) Use a large list of keywords
9) Link from sites with different LinkRanks
10) Monitor your keywords and adjust strategy as needed
11) Be patient waiting for results—they can occur anywhere from one to nine months
12) Get PR for your blog and website

After you have performed these steps, the main way to continue to improve your off-page optimization is to create tons of content and gain quality links to your site. While this is important, you will be nowhere unless you submit the content you create to directories and off-page SEOO

What Not to Do:

Don't use Link Farms, sites that collect major amounts of links and don't hire companies outside of your industry that promise to create links for you; bad links = less relevance points; good, organic links = more relevance points.

Don't use irrelevant keywords and misspellings—don't try to cover different spellings of the same term in different links—it is not viewed favorably by the bots.

Don't create links to spammy or trashy websites or blogs—as they get penalized by the bots, so will you if you are linked to them.

Directory Listings

A web directory is a directory on the Internet. It is usually linked to other sites and organizes those sites by category. Web directories are not search engines and are used by humans to locate the information they seek. You can submit your own site to a web directory for listing, and listing in as many web directories as possible is highly recommended. When possible, include your website or blog link, and your listing will turn into an inbound link to your content, which is a 'point' in the optimization game. Well known web directories include Yahoo! Directory and the Open Directory Project (ODP). ODP has an extensive categorization and large number of listings and its free availability for use by other directories and search engines.

The following web directory "characteristics" list was copied from Wikipedia (http://en.wikipedia.org/wiki/Web_directory):

- **Free submission**—there is no charge for the review and listing of the site
- **Reciprocal link**—a link back to the directory must be added somewhere on the submitted site in order to get listed in the directory
- **Paid submission**—a one-time or recurring fee is charged for reviewing/listing the submitted link
- **Nofollow**—there is a nofollow attribute associated with the link, meaning search engines will give no weight to the link
- **Featured listing**—the link is given a premium position in a category (or multiple categories) or other sections of the directory, such as the homepage. Sometimes called sponsored listing.
- **Bid for position**—where directory or other listing sites are ordered based on bids
- **Affiliate links**—where the directory earns commission for referred customers from the listed websites

Human-Edited Directories

These directories are created and maintained by editors selective in whom they include or link to within the directory. This category can include special-interest directories, such as a "doctor directory" or a "pharmacist directory." It is desirable to seek listing in human-edited directories, as the sources in these directories has a higher likelihood of being reputable, thus there is a better chance of getting a good link back from participating in that directory. Some directories use features that keep search engines from rating links, thus they are of no benefit to your search engine ranking (the nofollow attribute that we discussed earlier in Chapter 3). Human-edited directories are often targeted by professional search engine optimizers on the basis that links from reputable sources will improve rankings in the major search engines compared to non-human directories. Some directories may prevent search engines from rating a displayed link by using redirects, nofollow attributes, or other techniques.

Directory Tips

Directories exist for all types of content in every industry, so do some creative research to find out which directories exist within your market, profession, or area of particular interest. There are "article directories,"—directories that list articles within your field—as well as plenty of respected journals and academic institutions that participate in directories in your industry. Respected publications and academic institutions give better and more reputable links than other sources, so these are good directories to go after to help increase linking from high level sources, thus increasing your content's relevance. There might be thousands of these directories, but you really only want to list your content in the most popular fifty to one hundred of these. You can submit to thousands, but it takes time, and the value of the links from the lower-rated directories may not be worth it.

Advantages of Directory Article Submission:

- Multiple deep links allowed in your content in most directories that point back to your content
- The page with your article that points to your page is 100 percent related to your site's content
- Directory indexing occurs fast, so usually within a month the search engines will find your content and index it
- Other sites in your field will see your content and link to it, giving you a boost
- The article will generate visits to your site from interested parties in general
- Adding articles keeps you "active" in the eyes of the search bots

Creative Ways to Get Links from High Level Relevant Sources

Educational institutions, government institutions, corporations, major charities, and other institutions that serve your specific area of expertise are the best places to obtain links; their links offer the most value when it points to your website. So how do you get these sources to link to you? Start the way anyone would: create novel, interesting, high quality content within your specific area of expertise, create it consistently, and share it within the social Internet. Somewhere along the line, something you say is going to get the attention of someone at some type of institution mentioned above, and that person will link to you. You can also strategize to gain these links. For instance, universities often have forums, blogs, and job boards. Using the strategy of being "social," get involved in or start a conversation in a forum, reply to a blog, or post a job on a job board, making sure you incorporate a link to your content. There; done! You have just created a link with a .go, .edu, or high-level corporate .com and created major link power for yourself. Don't practice this randomly but with purpose. If you post high-quality content, or people appreciate the content you post, you may find people from the institution generate more links for you, and your high-level linking becomes more automatic and proliferates rapidly. Remember, always try to get the deeper pages of your website linked, and avoid creating links to your homepage or the major landing pages within your site.

Another method is to use a feature common to many portals, the "backlink data dump." For example, Yahoo! Site Explorer provides you the ability to get a "data dump" (i.e., a cache of information on the sites that link to your *competitors*). You can review this data to see if you too can get links from these sources, or find out which directories link to them, and submit your site as well so you can compete better. Or, if that is not possible, these may be places you may want to place paid ads so people may find you first.

Your best bet lies in content creation, so write content, post it, and try to get it published or syndicated; each step will lead to more and better links, giving you the "link juice" you need to stay ahead of your competitors.

Wikis

From Wikipedia (http://en.wikipedia.org/wiki/Wiki): A wiki (/wɪki/*WIK-ee*) is a website that allows the easy creation and editing of any number of interlinked web pages via a web browser. Wikis support multiple contributors with a shared responsibility for creating and maintaining content, typically focused around text and pictures. Wikis are typically powered by wiki software and are often used to create collaborative websites, to power community websites, for personal note taking, in corporate intranets, and in knowledge management systems. *Wikis* are simple web pages that groups, friends, and families can edit together.

Ward Cunningham and Bo Leuf, in their book *The Wiki Way: Quick Collaboration on the Web*, described the essence of the Wiki concept as follows:

- A wiki invites all users to edit any page or to create new pages within the wiki website, using only a plain-vanilla web browser without any extra add-ons.
- Wiki promotes meaningful topic associations between different pages by making page link creation almost intuitively easy and showing whether an intended target page exists or not.
- A wiki is not a carefully crafted site for casual visitors. Instead, it seeks to involve the visitor in an ongoing process of creation and collaboration that constantly changes the website landscape.

Further, the Wiki site explains:

- A wiki enables documents to be written collaboratively, in a simple markup language using a web browser. A single page in a wiki website is referred to as a "wiki page," whiles the entire collection of pages, which are usually well interconnected by hyperlinks, is "the wiki." A wiki is essentially a database for creating, browsing, and searching through information.
- A defining characteristic of wiki technology is the ease with which pages can be created and updated. Generally, there is no review before modifications are accepted. Many wikis are open to alteration by the general public without requiring them to register user accounts. Sometimes logging in for a session is recommended, to create a "wiki-signature" cookie for signing edits automatically. Many edits, however, can be made in real-time and appear almost instantly online. This can facilitate abuse of the system. Private wiki servers require user authentication to edit pages, and sometimes even to read them.

Wikis are usually published by an organization or person who maintains control over information that is submitted. Often, the wiki developer will reserve editorial rights, i.e., the ability to remove material that is offensive or "off topic." Wikis that accept any content without firm rules are called "open purpose" wikis.

One of the benefits of wikis is the ability of people to correct "mistakes." Wiki's have a recent-changes page, a list of recent edits that can be checked to verify the validity of recent content addition.

In healthcare and science, typically areas with high standards for publications, there is a newer kind of wiki gaining steam known as expert-moderated wikis. The system enables review via the peer review system, incorporating links to trusted versions of the article quoted or written in the wiki.

Commoncraft has an excellent video explaining wikis here: http://www.youtube.com/watch?v=-dnL00TdmLY. Creating a wiki is similar to putting yourself in the encyclopedia, or the "who's who" directory—so be sure to publicize your wiki and share it across your social media sites. Wikis are incredibly participatory and useful for building a community of your own around your content.

Paid-For Directories

Some web directories have a pay-for-inclusion model. These listings are generally posted quicker than non-paying directories, and there are generally fewer listings because of the fee. They offer features similar to the *Yellow Pages*, where your listing can be bold, you can have a banner or a photo, etc. If you pay, you should give preference to directories that do not support the nofollow attribute—remember, that is the attribute that makes your listing unsearchable by Googlebots.

Bid-for-Position Directories

These are directories that work similar to AdWords. You bid a certain amount, and if your bid is the highest, you rank highest on the directory. If one or more people bid higher than you, they appear ahead of you. Higher listings increase visibility, etc.

The following is a list of the most popular directories, according to Wikipedia:

- *AboutUs.org*—A wiki-based web directory.
- *Ansearch*—Web search and directories focusing on the US, UK, Australia, and New Zealand.
- *Best of the Web Directory*—Lists content-rich, well-designed websites categorized by topic and region.
- *JoeAnt*—A community of editors from the now-defunct Go.com volunteer-edited directory.
- *Open Directory Project (a.k.a. DMoz or ODP)*—The largest directory of the web. Its open content is mirrored at many sites, including the Google Directory.
- *Starting Point Directory*—A human-edited general directory organizing sites by category.
- *World Wide Web Virtual Library (VLIB)*—The oldest directory of the web.
- *Yahoo! Directory*—The first service that Yahoo! offered.
- *Biographicon*—A directory of biographical entries.
- *Business.com*—Business directory that charges a fee for review and operates as a pay-per-click search engine.
- *VFunk*—Online directory that specializes in listing and categorizing global dance music and urban lifestyle listings.

The major search engines, such as Yahoo!, Bing, Google and Google Buzz, and MSN, have directories as well. Be absolutely sure you have submitted your URL to these major players first.

Sitemaps

A sitemap is a list of the pages on your website. By creating a sitemap and submitting it to search engines, you are ensuring the search engines know about everything on your site, including pages that aren't able to be found by the engine (e.g., through the crawling process). The search engines aren't guaranteed to crawl every page on a website, but the engines will crawl your sitemap to study your site's architecture. The bots have a memory, so crawling the sitemap allows the engine to crawl your site better the next time it crawls. Information gleaned from sitemaps include when the site was last changed, how often it is updated, and how important each page is to the other pages on the site. The engines "learn" by crawling, and you want your website to be part of that learning process each time they crawl. You get points when your website changes, as the bots prefer updated sites to static ones.

Sitemaps can provide information on web pages, videos, news, Global Positioning Systems (GPS) and other web applications. Think of the Sitemap's role in your search engine optimization as a "hint" to the bots as to where to look and how the bot can do a better job indexing your site. Every time you update your website, you want to resubmit your sitemap URL to search engines so they can find your new stuff.

Search Engine Submission of Sitemaps

Sitemap Writer Pro is a great place to find tools that help you submit your sitemaps to search engines. Go to http://www.sitemapwriter.com/notify.php and easily cut and paste the URL of your website sitemap for immediate inclusion in all the major search engines, and you can go to http://www.sitemapwriter.com/notify.php?crawler=all&url=[sitemap_url] to submit your sitemap to all search engines that support XML sitemaps (a particular type of sitemap; the most common type).

URL Redirection Services

A redirect service offers links that redirect users to the desired website. The typical benefit to the user is the use of a memorable domain name, also called a top-level domain (TLD). For instance, someone who owns the domain www.optician.com might charge people with optician-related websites in different geographical locations who want that TLD and provide a redirect, so potential clients are led to believe your website emanates from a TLD, which adds credibility.

Manual Redirects

If you want to send someone from one of your pages to a newer or better version of a similar page, you can redirect them by putting a note on the page, something like, "Please visit our new webpage at http://www.youreyesite.com/new-web-page.htm." There are also automatic ways to do this.

HTTP Status Codes, 3xx Redirection

The following is taken from WikiPedia:

The client must take additional action to complete the request.

The HTTP standard defines several status codes for redirection:

- 300 multiple choices (e.g., offer different languages)
- 301 moved permanently
- 302 found (originally temporary redirect, but now commonly used to specify redirection for unspecified reason)
- 307 temporary redirect

A 301 (moved permanently) redirect is the best method for redirecting a webpage. A redirect occurs when you have more than one domain name, and you direct the other domain(s) towards the main domain. Redirecting is also sometimes called "domain forwarding," or "forwarding." Redirecting a page shouldn't harm its search engine ranking. The prefix 301 indicates something has been moved permanently. It is a commonly used method of making a website available under many URLs.

301 redirects are used when:

- You've changed your site's domain name and want to direct traffic to the new domain without losing visitors.
- People visit your site by going through different URLs.
- People might mistype your domain name. For instance, you might think you're lucky owning the domain ophthalmologist.com, but how many people searching for an ophthalmologist know how to spell it the right way? So you might own opthalmologist.com, opthamologist.com, and opthomologist.com. You would redirect the three misspellings to your actual website, ophthalmologist.com, to ensure more people found your domain. You might also consider purchasing different extensions (.net, .info, .org) for the different domains,
- You might be reserving "top-level domains" (TLDs)—buying domain names as an investment to sell later, and in the meantime you can "park" them – "parking" refers to activating them and pointing them to your site.
- Short aliases for long URLs and URL shortening; particularly useful for microblogging sites like Twitter, short URL aliases can be created

through services like tinyurl.com, which compact a long URL to something that takes up much less space. For practice, copy your URL (such as http://www.your.com). Visit http://www.tinyurl.com and paste your URL into the tool. It will generate a shortened URL that you can use as a reference to your site or blog when you tweet (which are limited to 140 characters, URLs included).

- If you have your page bookmarked yet desire to change the URL and don't want to lose the bookmark, you can use redirect to maintain it.
- Some people have used redirect to manipulate search engines and steer people away from their intended destination, trying to sell them something. Modern search engines can recognize this fraud and will penalize sites that use this method of redirect.

Some other methods of off-page optimization include:

Content restructure—if your content has the proper structure to it in terms of its URL, including relevant terms, it will be easier to find in directories by people searching those terms.

robots.txt file—these files instruct search engine bots as to which content is allowed and which is prohibited for crawling.

Google Analytics—allows you to refine the ability to track and react to user actions within the website. See Chapter 16, Discussing Analytics.

Search Queries and Operators (see next section)

Search Operators

Search operators are phrases entered in search engines with a technique that often includes symbols or additional text that gear your search toward a specific type of content, list, or file. We will discuss operators supported by the major search engines in a later chapter.

Advanced search operators (ASOs) let you organize your search, enabling you to find things that you can't with a standard keyword or key phrase search. You can also use them to analyze website optimization issues. Knowing how to research can give you a competitive advantage when competing for optimization.

To use an operator, enter the subject you want to search, followed by "operator." For instance, if you are searching fishing but want to modify the results, you can try the following:

1. [**-keyword**] excludes the keyword from the search results, e.g., [fishing - Montana] shows results for all types of fishing *except* fishing in Montana
2. [**+keyword**] allows for a *forced keyword inclusion* so with our example [fishing+Montana] would find all types of fishing in Montana
3. [**'key phrase'**] shows search results for the exact phrase, e.g., ['fishing montana'] and only that phrase
4. [**keyword1 OR keyword2**] shows results for *at least one* of the keywords, e.g., [fishing OR montana]

A list of Google advanced search operators is available at http://www.Google.com/intl/en/help/operators.html

A list of Yahoo! advanced search operators is available at http://www.bruceclay.com/newsletter/1004/seoperators.html

A list of Bing advanced search operators is available at http://www.bing.com/toolbox/blogs/search/archive/2005/06/24/432439.aspx

Other Engines and Query Sources

Google Blog Search allows for the following advanced search operators:

- **inblogtitle:***keyword* restricts search to blog titles
- **blogurl:***keyword* restricts search to blog URLs
- **inpostauthor:***Author Name* searches for posts written by a specified author
- *link:http://www.domain.com* returns blogs linking to the given page

Google News provides for these advanced search operators:

- **location:***country* or **location:***state* restricts results to news originating from the given location
- **source:***news source* returns news by the given news source
- *site:mysite.com* shows if your site is indexed by Google News

Technorati lets you explore blog back links ("reactions") via its *advanced search:* **http://technorati.com/blogs/www.blogdomain.com?reactions**

BlogPulse tracks blog mentions for any **URL** you specify. While being a good addition to a *technorati* blog reactions search, BlogPulse allows for better tracking options and also works for non-blog sites.

CHAPTER 5: LINKING

Internal Linking

Internal links refer to the links you create on your site that link to other pages. Yes, you can do this, and yes, it can help. It's actually a good tool to use to increase PageRank. Internal linking is sometimes referred to as "spidering." When properly spidered, you can get a relevance boost from these links. When you link from one page that contains search-relevant keywords or keyword phrases to another page with similar keywords and phrases, each page boosts the other. You can also use your more-popular pages to link to less-popular pages, giving the less-popular pages a PageRank boost just by associating them with the more-popular page.

The ultimate goal of internal linking is to ensure that every page of your website gets found by search engine bots. It's an incredibly easy way to boost your PageRank. The more pages you are credited with having on your website, the better PageRank the site gets through spidering. It also increases the number of search phrases your website is credited with. Remember, the more content you are credited with, the higher your PageRank.

Sites use what are called "navigation scripts" within their architecture that define the structure and links within a website. Bots read these scripts to learn what is going on, and then categorize and index it. "script-based" and/or "image-based" navigation systems are used to categorize your website. If your links are structured the wrong way, the navigation system may be hiding pages from the search bots. Since search bots aren't sophisticated enough to be able to catalog images, image-based navigation spidering doesn't really add much relevancy to your PageRank. Also, images can slow down uploads of your content, causing people to leave your site earlier than they normally might have, so there is more than one reason not to pepper your content with too many images.

Another benefit of internal linking is that the "closer" a page is to your homepage—i.e., the fewer links it takes to get from a home page or popular page to another—the better the ranking the less-popular page will have. Please *do not* link all your pages to your home page! This can add to the confusion of the home page and take relevance points away from each individual link. Be picky when you pick the links on your homepage, and use pages that you really want to drive traffic to but are too narrowly focused to be main pages or major landing pages.

Ways to Improve Your Internal Links for Optimization:

a. Add two to five contextual links to each article page
b. Use relevant terms as anchor text when creating contextual links
c. Never link to more than fifty other pages from any one page
   1. If PageRank (PR) is lower than four, never link to more than twenty-five other pages from that page
d. Do not add the same links to every page of the site
e. Try not to link from one page to another multiple times
f. Follow link equity when linking to the most "in-need" pages in order to pass the highest level of link juice found in Appendix B
g. Make sure your site contains *no broken links*
h. There is a tool by Xenu called LinkSleuth—it can scan your site for broken links. Search engines dislike sites with broken links, so identify them, and then fix them or get rid of them.

External Linking (Building Back Links): Gaining Links Targeting Specific Pages and Key Phrases

External links are links to web pages that exist outside your web page. An external link "points" to another web page, and when it is clicked, you're able to go there. There is some debate over this, but many professional SEOs believe that external links are the most powerful source of ranking power. The more the better. The reason SEOs believe so much weight is given to this metric is that it is the hardest thing to fake—people either link to you because they like your content or they don't. External links provide relevancy clues that are important for search engines to do a good job ranking you. The anchor text in these links is usually highly reflective of the page being linked to, containing keywords and key phrases relevant to the content you produce.

Search engines also keep detailed data when they discover the existence of a new link or the disappearance of a link. You appear more relevant when the data on your website changes regularly, so creating links can help.

Methods of Gaining External Links

Turn brand mentions into links. A great way to use preexisting content in order to acquire traffic-yielding links. This consists of finding blogs and websites that already mention you but may not link to the pages being targeted (or don't link to the correct place). You may want to send the webmaster an e-mail requesting a link as they already refer to you or your business.

Contact newspapers and local websites or blogs. Many major news syndicates feed off small community sources. What may begin as a few small links can turn into links from major conglomerates like MSNBC or Yahoo! News.

Send e-mails to site owners requesting a non-reciprocal link. While this is a more challenging way of getting an external link, as you are getting and not giving, you may earn a few excellent links this way.

Social bookmarking can be used effectively to create links to articles (see Social Bookmarking, Chapter 10).

One of the more famous papers on the topic of building external links was written by Apostolos Gerasoulis and others at Rutgers University on applying link analysis to web search: http://www.case.lehigh.edu/~brian/pubs/1999/www8/

The News Hook

If you are posting interesting and helpful content or you are early to get news and information to your industry online, the likelihood other people will find it compelling may cause them to link to you—you can "hook" people into linking to you by posting the latest and greatest. This is a powerful method of link building. These are particularly strong links, and they go a long way toward reputation enhancement. Being timely "to print" with hot industry topics helps you by increasing the perception that you are an important participant in the community and encourages people to link to you.

Other Hooks that Might Draw Links

1. Posting lists of resources
2. Offering free stuff—people *love* free stuff!
3. Contests that enable people to show off their smarts, awards for participation or asking for user input.
4. Create a post that evokes passion over a cause
5. Humorous posts

Domain Authority and Trust

Authority, when referring to web content, gauges its importance. Authority of a web page is based on the depth and quality of its external links. The Google bots look at who links to you and determines whether you are a high-level source of information for the keyword or keyword phrase being searched. For instance, a highly authoritative domain for someone interested in collegiate sports to link to would be the National Collegiate Athletic Association (NCAA) website. This is a site that is highly linked in sports at the collegiate level, and by connecting to it, your content becomes part of this well respected and highly regarded matrix. On the contrary, if it links to a website offering, for example, some spammy college travel agency, it is not likely you are linking to a high-authority website, and this can bring down the authority of yours. The higher authority the content has, the more trust is established.

Once you have identified which top-level domains (TLDs) you want to get links from, you should proceed to attempt to get links from a large number of those domains. Don't forget your lower-level but highly respected pages! Direct links to lower-level pages create anchor text opportunities for those pages as well. Each major theme or topic of the site should be the subject of its own link campaign. A great way to obtain links is in your back yard, starting with local businesses. Begin in your office building by asking your neighbors and vendors. For example, a link from the local chamber of commerce is likely to be helpful in ranking for local search terms.

Anchor Text and Linking

Anchor text (see section on anchor text, Chapter 6) has been rated by leading SEOs as one of the most powerful factors in search rankings. Anchor text refers to keyword or keyword phrases placed within the architecture of a particular page.

If a site has a page selling kitchen sinks, and the page has numerous links pointing to it that use the anchor text "kitchen sinks," the anchor text is likely to help the page's rankings for related search terms. In other words, somebody else's keyword or phrase, being the same as the ones you use, and linked to your page, gives you a boost in relevance.

Syndicating Content: An Effective Link-Building Strategy

You may also choose to try to get your content placed on another site (e.g., syndicating it). One reason for this is to provide the content for another site in return for a link to their site. If you pursue this strategy, consider implementing unique new content for syndication instead of syndicating articles from your site. Having others publish articles from your site will result in the engines seeing your article as duplicate content, and this can have some undesirable affects.

Directories can be a great way to obtain links too. There are a large number of directories out there, and they may or may not require money in order to obtain a listing. (See the section on directories in Chapter 7.) A comprehensive list of directories is available from *Strongest Links* at

http://www.strongestlinks.com/directories.php

Social Media Sites and Your Link-Building Campaign

Social media bookmarking sites, such as StumbleUpon, can play a big role in a link-building campaign. Becoming "popular" on these sites can bring in a tremendous amount of traffic and links. While social news sites like Digg and Reddit bring lots of traffic, this traffic is usually of low quality and will have a low revenue impact on the site receiving it. The real ballgame is to get the links that result in high revenue impact. For example, stories that make it to the home page of Digg can receive tens of thousands of visitors and hundreds of links. While many of these links are transient in nature, there is also a significant number of more permanent, high-quality links that result.

Better still, articles covering topics that happen to relate to competitive keywords can end up ranking very well, very quickly, for those very competitive keywords. The key insight into how to make that happen is to use the competitive keyword in the title of the article itself and in the title of the article submission.

Getting Searchial Links: Links From Social Media Profiles

Some social media sites, such as LinkedIn, allow you to link back to your own sites in your personal profile, and these links are not NoFollowed (meaning they pass link juice). Leveraging this can be a great tactic, as it is simple and immediate.

In the case of LinkedIn, for example, the process takes a few steps:

- Login to LinkedIn.
- Click Account & Settings in the top right corner of the LinkedIn screen.
- Click on My Profile to edit it.
- Click Websites to edit your websites.
- This will present you with the ability to edit your additional information (click edit).
- Add a website listing. Start with the box that says Choose and select Other. This is what will allow you to specify keyword-rich anchor text.
- Then enter in the correct URL and click Save Changes.
- Next we have to make your websites visible to the public (and the search engines).
- On the upper right-hand side of your screen, you will see Edit Public Profile Settings. Click on it.
- On the next screen, under Public Profile, make sure "websites" is checked. This is what tells LinkedIn to display that publicly.
- Click Save Changes.

Note that the above process was what was required as of early 2009, but the specifics may evolve over time as LinkedIn releases updates.

Eighteen other social media communities that do not use nofollow links in their public profiles at the time of publication are:

1. Flickr
2. Digg
3. Propeller
4. Technorati
5. MyBlogLog
6. BloggingZoom
7. Current
8. Kirtsy
9. PostOnFire
10. CoRank
11. Fark
12. Slashdot
13. Metafilter
14. Mixx
15. Hugg
16. Sk*rt
17. Stirr'dup
18. Linkinn

By the time this book is published, it is probable that some or most of these directories will have started using the nofollow attribute, but for now they are all dofollow. Remember, a site that uses the nofollow rule doesn't let the Google bot crawl your site and index the content, so you don't get value by being bumped up in search engines for the keywords and key phrases you include in content. Dofollow sites give you that bump. It's a good idea to create links to your site or blog from all of these sites, and a great place to get started with your linking efforts.

People often create links in their content to relevant Wikipedia pages, as Wikipedia references usually come up in the top page, if not the top position for many keyword searches. However, Wikipedia uses the nofollow attribute, so creating a link to Wikipedia, as it seems like a high-level link that might provide a lot of link juice is futile because of the nofollow attribute.

Backlinking for Links

Backlinking, or seeing who links to whom, is one of the oldest methods of putting together a link campaign. If someone links to one publisher's competitor, there is a decent chance that he or she might be willing to link to the other publisher as well. The process starts by identifying sites related to the market space of the publisher, including directly competitive and non-competitive authoritative sites. Then, using tools like the "linkdomain: command" in Yahoo! Search or Linkscape, you obtain a list of the sites linking to those sites. The precise form of the advanced search operator command in Yahoo! is: linkdomain: targetdomain.com-site:targetdomain.com. The purpose of including the "-site:targetdomain.com" at the end is to exclude links from the site back to itself. However, the output from Yahoo! Search is not easily downloaded. Third party tools, such as SEOmoz's Linkscape or Link Diagnosis, can provide you backlink data in a spreadsheet format. Note, however, that if you enter the simpler form of the command, linkdomain:targetdomain.com, this will bring you into Yahoo! Site Explorer.

Helpful Search Terms, Phrases, and Advanced Query Parameters

Search around, and you will find a plethora of tools in search engines that enable link-related research. The following is a list of commands and links you can dig into to learn more about who links to whom, which keywords competitors are using, where they seek links from, and a ton of other useful information that can help you with your searchial efforts when determining which non-obvious keywords and key phrases might be used to drive search to your content.

inanchor:keyword—Allows you to determine a particular page's relevance based on the quality and quantity of links pointing to it. For example, if you type into a search bar the phrase inanchor:editing, it will return the pages that have the best inbound links that include the word editing in the anchor text.

Taking this further, you can look at pages on a particular website that has the best anchor text for a particular keyword by using: inanchor:keyword site:domain-to-check.com

This search can diagnose whether a site has "unnatural" links. Unnatural links are links that are purchased. Google frowns upon this practice and has been known to de-index sites that buy or sell links. If 80 percent or more of the site's links include the major keyword in their market, it's likely they are buying links or doing something to steer the anchor text.

This operator is *very* valuable as a tool to spy on a competitor's site to understand its relative strength competing for a keyword.

There are other related operators as well:

Intext:keyword—shows pages that have the phrase implemented in their text
Intitle:keyword—shows pages that have the phrase in the page title
Inurl:keyword—shows pages that have the phrase in the URL for the page

linkdomain:http://www.domain-to-check.com/ -site:domain-to-check.com keyword

Find keyword-related sites that link to a specific domain. This command needs to be executed in Yahoo! Search, since Yahoo! is the only search engine that supports the linkdomain command. This operator is valuable in learning about a competitor's site and its strength in competing for any given keyword.

linkdomain:domain-to-check.com site:.edu 'keyword'

This command is useful in finding the .edu sites that link to a competitor's domain, where the site uses a target keyword. This command also needs to be executed in Yahoo! Search. Take a competitor's site and use "resources" for the keyword, and see what happens. Remember, .edu is a TLD (top-level domain) that has particular value in getting links from.

yourdomain.com -site:yourdomain.com, with the &as_qdr parameter

This one is a bit trickier. In Google, perform a search on domain-to-check.com ñsite:domain-to-check.com. Then add &as_qdr=d to the end of the URL on the search results page and reload the page. This will show the mentions that domain-to-check.com has received in the past twenty-four hours.

linkdomain:competitor1.com linkdomain:competitor2.com -linkdomain:yourdomain.com

This command shows you who links to two of your competitors but not to you. Sites that link to more than one competitor, but not to you, may well be interested in learning about another site providing information on similar topics, particularly if your site does one or more things better than the sites they already link to.

intext:domain-to-check.com

This command can help you rapidly identify sites that reference domain-to-check.com but don't implement that reference as a link. This command is powerful because it can be used to identify "lost" links to your domain. By identifying sites that reference your domain, you can contact those sites and ask them to convert it into a link. Since these are sites that already endorse your site (most likely they are, at any rate), the conversion rate can be expected to be reasonably high.

Sources and Additional Resources

Here are some additional resources for learning about search operators and how they can help with a variety of tasks:

- Advanced Link Operation article on Search Engine Journal: http://www.searchenginejournal.com/advanced-link-operator-to-explore-your-competitors-back links/6966/
- Getting links from known, quality linkers - http://searchengineland.com/080710-161019.php
- Linkscape
- Advanced tools exist to do much of the work for the publisher. One example is SEOmoz's Linkscape. Linkscape is a tool that has been developed based on a crawl of the web conducted by SEOmoz, as well as a variety of parties engaged by SEOmoz. You can see a complete list of Linkscape's sources here. http://www.seomoz.org/linkscape/help/sources

CHAPTER 6:
BLOGGING AND CONTENT CREATION

Not too long ago, most of the information and news we received was filtered through major media organizations controlled by a few wealthy individuals or corporations. We heard what they wanted us to hear based on what they chose to publish or not. Their opinions became the opinions of large portions of society, and their power increased as the popularity of their periodical or TV show increased. Information people were exposed to was heavily biased, and an individual had no way to get their message to a large audience, while media outlets historically provide and censor news to suit their bias. *The New York Times* is often described as having a liberal bent and Fox News as being conservative.

Blogging has changed all that.

A blog is basically a publishing tool. Blogs, or "web logs," have created a new media, where anyone and everyone can publish and immediately receive exposure of their content to a worldwide audience. We've become our own editors, and there are no barriers to what we can or cannot say. The power of this new media is awesome, not only because our opinion can be heard, but because we have the ability to filter through a ton of media posted by other bloggers and are no longer limited in what information we take in through just a few major media outlets. Blogging has created competition for the mass media and tempered its ability to influence events that occur on the world stage; it has put the power of the media into the hands of the individual for possibly the first time ever.

Your blog will be important to your business-marketing efforts for four reasons:

1. **It enables you to build and strengthen your relationships**—Blogging is about content. It gives you the ability to update content and provide information to people regularly.
2. **It will help raise your profile**—Frequent updates to relevant information is recognized by major search engines and will help you get ranked higher in search results.
3. **It raises your relevance within social media suites**—The more you utilize sites like LinkedIn, Facebook, Twitter, and others for your business, the higher you'll rank in major search engines, and for the suites that are nofollow in relationship to the general Internet, you still raise your rank within most of these sites through these methods. It's searchial media marketing.*
4. **It can establish relevant links to and from other Internet sites**—Linking in your blog is the most important activity you can undertake for your efforts to gain relevance and show up more prominently in search engines.

*Just a reminder—the major purpose of your social media efforts is to appear more relevant to the "conversation" happening on the Internet; relevance is rewarded by higher search engine rankings, so while your efforts may draw few people directly into your establishment, those searching for your particular type of business will be more likely to find you if you are ranked higher for speific keywords or key phrases. Remember, when you blog, you are not blogging to an audience—that is the wrong attitude. You are blogging to add your voice to the conversation. So blog, read blogs, and reply to blogs. Interact on forums and discussion boards and link to your blog as well. Take part, don't lecture, but use the keywords and key phrases you think people are using to search for your services, and you'll get more of the searchial effect.

If a blog is essentially a website, why blog? The answer is because blogging provides a way to post information, opinions, or reviews with regularity without having to hire anyone to post it for you. It is your online journal. If you want to add text to your website, unless you know how to write code, you likely have to pay a webmaster to code and redesign the face of your website to add it and make it look nice. With a blog, you can post commentary, graphics, images, discussions, news, or anything you want much more easily. Bots notice and favor updated information; a blog is a way to constantly add content to the conversation and get your keywords and key phrases out there, keeping them fresh. Additionally, readers are able to leave interactive commentary about what they have read. Remember, when you add, change, or post new content to your website or blog, the bots and spiders that rank you give you points for having something "new." It's much easier to do this regularly on a blog than a website.

The Society for New Communications Research (SNCR) issued a report about new media (albeit *way* back in 2007). In it, they polled organizations to find out which new media tools were the most frequently implemented. Blogs won out by a large margin, but 56 percent of responders reported using a social network as well.

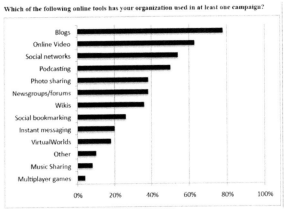

from http://www.marketingvox.com/social-media-marketing-still-lacks-strong-metrics-037724/)
credit to Society for New Communications Research http://sncr.org/

Blogging takes place through web services known as content management systems, or CMS, a fancy way of describing a page that you can intuitively and easily edit whenever you want, with whatever you want, without using a webmaster or having any knowledge of how to HTML code. There are many CMSs out there, such as Blogger, Wordpress, Yahoo! blogs, etc. Each system is slightly different from the others, but most let you easily post content and change it at will.

To establish your blog, I recommend you have your blog "hosted", or housed by a professional web hosting company; don't use a free content management system. I started my blog on Wordpress.com a long time ago, and it has taken on a life of its own, but I recently switched to a private domain via a paid host at Network Solutions. NetworkSolutions is the pinnacle of hosting companies, with excellent customer service and advanced development and hosting options. I highly recommend purchasing your domains from NetworkSolution and hiring them for other development needs. If you do want to use a free content management system, I have been told it is best to start with Blogger, which is owned by Google, and in theory may get preferential search engine listings compared to other content management systems. (Although I cannot verify this is true.) By owning your own domain to your blog, you build Internet intellectual property that has actual value. For long-term searchial benefit, it is best to have two domains out there where you build content and link to your website and blog.

On that note, if you have a blog that is on a separate domain than your website, it should be linked directly to the homepage of your website with a prominent button that enables people to easily navigate to it. If you have a blog through the same domain as your website, there is immediate searchial benefit over having a separately hosted blog, but in the long term, your searchial efforts are best served by developing two separate domains—blog and website—within the same content space. By regularly updating your blog and having it linked to your website, "update juice" passes through to your website; Google Search "sees" that the content on your website is regularly updated and will assign it a higher "grade" than a website that is not regularly updated. The result is a boost to your website in the search engine rankings. Now you're being searchial. My blog is youreyesite.net, and my website is youreyesite.com, and I choose to host them separately yet link them together to maximize my long-term searchial efforts while building my youreyesite brand and increasing the value of the domain IP concurrently.

How to use Your Blog

Putting something on your blog is referred to as "posting." I recommend posting new information on your blog at least three times a week. Typically, blog authors compose their information in a web-based interface built into a blogging system. The information is usually anything from a simple paragraph that discusses new research findings on diabetes, or your opinion on politics to a long, detailed blog post on whatever topic you decide to cover (although it is considered good etiquette to keep your blog posts as succinct as possible and keep the theme relatively consistent from post to post).

Make sure you are writing to a "community" and not an "audience"—audience is non-participatory; your blog will become the center of your searchial marketing effort; the "ial" in searchial derives from the ial in social (media) for a reason. Writing for an audience will limit reach and will keep you from retaining loyal followers and earning feeds. When you create a community, you create relationships that will benefit your online efforts, help increase relevance, and find a cache of loyal followers who will become clients.

Try to post innovative, interesting, and novel content. This will enable you to stand out from the pack and differentiate yourself. Content is one way to do this; other ways include using video, interviews, news tips, "tips of the day," or efforts that create social goodwill, such as charities. Your content should talk *to* your reader, not at them. Asking open-ended questions is one way to appear social in your media. This is the stuff people find interesting, not just cutting and pasting latest news stories that are relevant to your content. Once you do this, you will see your visitors spike.

As your blog increases in popularity, your readers will expect you to show up, and when you don't, they'll drop out. If you take too much time off, you may achieve a drop in your relevance and search engine rankings as others posting keywords and key phrases in your industry will gain ground on you.

Take your content, create an interesting or compelling tag line—I call it a "teaser"—and share it with a link to your blog and your Facebook and Twitter pages. This is an act that should be repeated with each post; you can also use the teasers on the social bookmarking sites. This form of indexing will quickly be recognized by search engines and increase your relevance. The teasers should be short, relevant headlines that grab a potential reader's attention. With a Twitter link, you get true searchial benefit; people leave via your link, find you, and Google indexes your tweets so others can find you. Twitter is dofollow and as a result is very searchial.

Make responding to your comments and followers a priority. A secret to searchial media success is to engage your readers and followers. This single act carries great weight in social media. If someone re-tweets (RT) your Twitter post, send him or her a simple "thank you for the RT." These gestures are appreciated more than you might realize. Take the time to answer questions. I can think of three instances right off the bat where answering a question from a reader of my blog led directly to an interview by a major media outlet. It doesn't happen all the time, but you'll see what I mean if you stick with it.

Give lots of credit and love to others; it's a great way to build a lot of long-term followers. Get creative! I had the idea to create the hashtag #MedicalMonday. The idea was to promote my web efforts and get recognized by appreciating the efforts of other in the "twittersphere." The idea was to have people acknowledge Twitter users who tweet on medical and healthcare issues on Monday, similar to how people acknowledge other Twitter users on Fridays with the #FollowFriday hashtag. I expanded the idea by creating an "award," the #MedicalMonday Healthcare "Twitizen" of the Week. On Mondays, I picked ten people tweeting interesting healthcare information and "nominated them for the award. The nomination was posted on the blog, Facebook, and tweeted to thousands of people. On Wednesdays, I would pick someone to win the award. The award became another blog post on Wednesdays, and I gave the award recipients the opportunity to write a guest post on my blog and promote their efforts or anything else they want. I have built tons of goodwill and gained hundreds of followers on Twitter through this effort, and its popularity is growing. I am now preparing a website "badge" that nominees and winners can display on their page, and when they do, the badge creates an inbound link to my blog, increasing the relevance of my blog. The likelihood of them taking that badge down is small, as people will use the award to further promote their efforts. The #MedicalMonday nominees who communicate with me or re-tweet their nomination or award are useful in the sense that I can contact them in the future asking for links, guest posts, or other content that might be useful, for even more re-tweeting of content I have yet to post. They become a network within a network for me, and as many are healthcare professionals, what they have to say is relevant to the keywords I use.

When I mention someone in my blog, I make sure I let him or her know about it, but I never ask for anything in return. It's one of those things you just don't do—call it social media etiquette if you will, but it won't get you anywhere to ask for anything. It's really a pay–it-back scenario on the social web, and your good deeds will eventually be rewarded.

Use guest poststers—it's tough to constantly come up with content, but you don't have to. Approach others, and make the offer to post on your blog seem too good to be true. If people don't bite, offer them a link to their website. If they don't understand the value of that, explain it to them!

Running along the sides of the blog page are often links to other pertinent or non-pertinent Internet sites—possibly lists of what you have posted on Facebook or Twitter. This is called the "blogroll." Blog posts are listed chronologically, placing the newest on top, making them more visible and accessible to visitors. The older articles are archived. A new post can be an interesting photo (I recently posted photos of blue-eyed celebrities and had my most hits in a day so far), or just a link to some press you received. It can be a video too. People seeking information through these channels often prefer not to be marketed to, so you should post informative, educational, or interesting news posts and refrain from posting about your sale or self-promoting your business. You are promoting your business by having a blog; links to your website and information, like your business name, address, and phone number should be easily visible, so people who visit it will appreciate the content you post and seek you out if they want. They are more likely to do so if your blog has a professional appeal without a marketing bent. Of course, every once in a while it is okay to plug yourself or a promotion you are having, but refrain from doing that too often or you will lose visitors. Search engines will pick up your blog, and I suggest you list your blog on blog search engines, such as Bloglines, BlogScope, and Technorati, to increase your visibility in the blogosphere.

Be sure to include plugin widgets ("share this" or other branded buttons). These buttons allow you to quickly and easily share your posts on Facebook, Twitter, and other social networking sites, a great way to proliferate your work and get followers, hence extra relevance. Most blogs offer helpful widgets that are easy to upload to your blog.

In searchial marketing, it is probable that your blog will become the centerpiece of your campaign and the vehicle with which you promote on Facebook, Twitter, LinkedIn, and elsewhere. I will describe how this is done after I post about Facebook and Twitter. As a blogger, you will develop your own media persona, and as a result, your blog will acquire it's own character as you find your inner author. It will build slowly at first—don't be in a rush to have more than fifty thousand followers overnight—but it will happen. In the beginning, get a feel for being a publisher. I found it difficult at first to find my blogging voice, but after a while, I started to realize I was in control of what I published, and there were not only no limits to what I could post on my blog, there was no one telling me how or what to post. That feeling empowered my marketing efforts and still does.

The popular book on social marketing *Groundswell*, lists these excellent tips for successful blogging:

1. *Start by listening*—A small knot of people at a cocktail party are conversing. Would you walk up to them and just start talking? Or would you listen first, and see how you might join their conversation? The blogosphere is the same. Listen to what's being said out there before you dive in. Monitor the blogs in your industry, from competitors, pundits, and other influencers. For a more comprehensive view, hire a brand-monitoring service like Nielsen Buzzmetrics or TNS Cymfony.
2. *Determine a goal for the blog*—Will you focus on announcing new products? Supporting existing customers? Responding to news stories? Making your executives seem more human? Choose goals so you know where you are going.
3. *Estimate the return on investment (ROI)*—Using a spreadsheet, determine how you think the blog will pay off and what it will cost. This is especially helpful in gaining buy-in from other functions throughout the company and in disciplining your thinking.
4. *Develop a plan*—Some blogs have one author, others feature several. You'll also need to determine whether you'll have a single company blog or a policy that enables blogging to spread to many employees and many different blogs.
5. *Rehearse*—Write five or ten posts before allowing them to go live. This is your spring training—when you find out what it's going to be like without all the flashbulbs going off. It also allows you to explore the sort of topics you'll cover. If you can't write five practice posts, you're not ready for the big leagues.
6. *Develop an editorial process*—Who, if anyone, needs to review posts? (The general counsel of your company? The Chief Marketing Officer? A copy editor?) Who are your backup if these people aren't available? This process needs to be built lightweight for speed because you'll sometimes want to post quickly to respond to events and news items.
7. *Design the blog and its connection to your site.* You'll have to decide how—and even whether—to feature your blog on your company's home page, depending on how central you'd like it to be to the company's image. Your design and the way you link the blog to your site will communicate just how official this point-of-view is.
8. *Develop a marketing plan so people can find the blog*—Start with

traditional methods: e.g., a press release to get coverage from trade magazines in your area, and e-mails to your customers introducing the blog. You may also want to buy words on search engines. But remember that the blogosphere is a conversation—you're talking with people, not shouting at them. You can leverage the traffic of the popular blogs you identified in step 1: include links to those blogs in your posts, and post comments on them to lead people back to you. The text of your posts will also help—by using the names of your company and your products in the titles and text of posts, you will make it easier for people to find your blog in search engines.

9. *Remember, blogging is more than writing*—To be a successful blogger, you should start by monitoring the blogosphere and responding to what else is out there, not behaving as if you are in a vacuum. And remember that your blog will have comments—if it doesn't, there's no dialogue, and you're no longer talking with people as they make decisions about your products (and services)—and that's the whole point. Finally, many corporate blogs use moderation, vetting comments to make sure offensive and off-topic chatter doesn't mar the blog. This takes time too, but you should do it. You can delegate the tasks of monitoring other blogs and responding to and moderating comments, but someone has to do them, or your blog won't be part of the dialogue.

10. *Be honest.* People expect a blog to be a genuine statement of a person's opinion. This doesn't mean you can't be positive about your company, but you need to respond as a real person. Sometimes bad things happen to good people and good companies—like Dell's laptop batteries catching fire. Dell's first post on the topic actually linked to a picture of the "flaming notebook" and included this frank admission: "We ... are still investigating the cause." This was followed by posts about how to get defective batteries replaced, once the company had decided to offer replacements.[9] A company that responds honestly, even when things go wrong, boosts its credibility.

Chris Brogan, an Internet marketing thought leader, offers the following suggestions in his book *Social Media 101* (pg. 19), a book I highly recommend:

- *Consider placing your picture on the main page*
- *Make your About page robust*
- *Make it easy to contact you*
- *Consider what you talk about in your blog*

Blogging and Linking

Link building is the single most important thing to do with your blog to get searchial. For now, the simplified version goes like this: when another website or blog references your website or blog in the form of a link, you score a point in the relevance search. The more websites and blogs that recognize you as a source via a link, the more relevant your effort becomes on the Internet and the higher you get ranked in Google or other search engines.

Who links to you is important as well. A link from *Writers Digest* to an author's blog gets more points than a link from a non-related or unpopular website, demonstrating why you want sites with high-level reputations to link to you. You get links back by linking to blogs and/or websites that are highly relevant to the content you are creating and hoping your work gets recognized. You create these outbound links by simply clicking the appropriate button on your content management site, and you can even use certain websites to help you determine whom you should be linking to in order to help increase your relevancy. When you link to hundreds of other sites, some webmasters will recognize that you linked to them, will visit your site, and some will like your content enough to link back to you. That is how you grow your link profile. Additionally, blogs are ranked by Technorati and other blog search engines based on the number of incoming links.

Blogging and Tagging

Keywords and key phrases people search for become tags when you input them in a special box within the "dashboard" of your blog-management system. These become the "classification" words and phrases, and bots looking for content with those particular words or phrases can find your content easier when it is tagged for them. When people search for a dermatologist in "Rockville" and "Maryland," each word should be tagged; chains of tags are searched when put in the Google search bar. Google, in turn, pulls up the most relevant sites or blogs to the particular search tags entered. Search engine relevance builds from linking and from social media relevance. If you are writing about macular degeneration, you want to post the tags "macular" and "degeneration." You also want to post terms that appear in your post. Your post will do more for you if you can also get the name of your town and practice in the same post, and type tags for "macular, degeneration, Rockville and Maryland" in my case.

This is where blogging for searchial benefit becomes an art. You need to find ways to include the terms that are relevant to the people finding you in your posts (which often have very little to do with your practice), and then create the appropriate tags and links within your post. This is how your blog post gets searchial, and eventually floats higher in search results compared to people with poorer content, poorer linking, and poorer tag juice—the more posts you have written where the tags are specific and get linked to, the more recognition from the search engines, and the more likely it is to result in a higher ranking for you. This results in more recognition from people in your geographical region when they search for a particular topic you've blogged about.

I suggest you blog hardcore material with best content on Tuesdays and Wednesdays and then tone it down over the rest of the week to maintain an easy workload. Fridays, into the weekend, are the least read, so don't put effort in then. You'll notice spikes of visitors on days or weeks you didn't think you really did much. The reason for this is, these are the days or weeks after the Google bots have scoured the web and refreshed Google's index of website and blog posts. As you want your content read on both coasts, I have read suggestions that it benefits you to post blogs between the hours of 11:00 AM and 1:00 PM EDT.

Be Careful

Blogging can result in a range of legal liabilities and other unforeseen consequences, so be sure to post original, non-defamatory content to avoid legal issues. The rules for legal action online are still being written, but better not to step on toes. Be sure not to defame anyone or say anything that might be construed as libelous.

Getting Started

So you want to blog, but you're not sure you have enough to say? Just think of every encounter, or customer, in my case patient, as a potential post (of course you don't want to identify them in any way, shape, or form).

Every time you walk into the office or assist a customer over the phone, you may not encounter a problem, but you encounter a different personality dealing with the same problems. How do they report the problems to you? Via blog post. How do they deal with the problems? Blog post. How do you handle the problem specifically with their needs in mind? Blog post. When you see something rare—blog post. When you see something funny—blog post. When they say something funny—blog post. Get the idea?

A few years ago, I started the hashtag #badvisiondecision on Twitter. The idea came to me because I'd say in four or more out of ten patient encounters, patients reported engaging in a habit detrimental to their ocular health, be it over-wearing their contact lenses, smoking, eye rubbing, etc. After seeing the patient, I would microblog the poor behavior, bringing it to the attention of people who may not know better and, in the process, hopefully helping someone out there. These microblogs, including the #badvisiondecision hashtag, became popular, and many people now use them to raise public awareness of things that may lead to poor eye health.

Blogging involves staying on your toes and paying attention to events around you that may not normally grab your attention. I have to view things from the perspective of an author as well as that of a small business owner. It also involves developing the skill of expressing ideas in writing you previously would have glossed over, not having any reason to pen it. Authors seeking material should keep their eyes and minds open to what is happening around them, and when they see something that might qualify as material, they should try to come up with a creative angle, or a way to make the mundane more interesting by adding personal perspective.

An incredibly useful tool I use regularly to assist my blogging effort is Scribe. This is one of the best tools I have used, and with the site socialoomph, my proliferation efforts are greatly enhanced. With Scribe, you author the content, input it into their software, where it analyzes the headline you choose, makes suggestions to optimize your content, suggestions for keywords you may want to include, and grades your content so you can see where there is room for improvement so you can change it prior to publishing. The power of this tool is awesome, and I recommend it for beginning through advanced bloggers. Visit http://wp.me/pxAmm-zu and click on the Scribe link to learn more.

To get started with Wordpress, visit http://codex.Wordpress.org/Getting_Started_with_Wordpress. I find Wordpress to be one of the easiest and most intuitive of the blogging content-management systems. Other free content management systems you can try are Blogger, LiveJournal, OpenDiary, TypePad, Vox, ExpressionEngine, and Xanga.

Widgets

The term "widget" is short for "window gadget." A widget is a programming element that usually appears as some kind of box, badge, or button on a website. The widget is programmed to show information that can be interacted with to manipulate data when clicked on. The widgets you might commonly encounter may seem like "virtual" buttons that can be clicked with the cursor. Desktop widgets are used to provide easy access to programs you obtain with a click of the mouse. Your desktop clock and calendar are examples of widgets.

You can find widgets that are useful to your efforts by searching widgets in Google, and the content management system you host your blog on has widgets that are easily placed on your blog page. You want to populate your blog with widgets for the most common social media sites, such as Facebook and Twitter. Widgets include visitor counters, weather logs, search bars, image-sharing sites like Flickr, RSS widgets, blog stats, and more.

You want to choose widgets that benefit your efforts by providing helpful information for your visitors; making your site more interesting (e.g., interactive) to visitors; promoting your social media efforts; and making your site more navigable. You can also choose widgets that, when applied to your site, promise to link back to you in return. This creates an inbound link that also helps improve your rank in search engines.

Really Simple Syndication (RSS) is a program that allows anyone reading your content (usually on your blog) to "subscribe" to it. Someone interested in the content you create clicks on the RSS widget (logo) within your blog, and your content is "fed" into their "feed"; in essence, they are subscribing to you.

If they use Google Reader and have News, their YouTube and, for example, three blogs they subscribe to all show up when they open their reader. If they like what you say and subscribe to you by clicking the orange RSS icon, your content will show up in their feed as well. One of the goals of blogging for your business is to subscribe as many RSS feeds as possible; it is a way to "lock" people into your content, and it creates a relevance link for you. Looking at it another way, the amount of people who subscribe to your content is a good metric of your blog's popularity.

It's not enough to have an RSS widget on your website. You have to make subscribing easy for your visitors. There are many ways people should be able to subscribe to your content. Not everybody uses a reader; many people use e-mail subscriptions, where blogs show up in their inbox. Enable e-mail of your blog a benefit for people who mostly use e-mail, and if it's video, provide the links on your blog or by e-mail. I recommend the tool Feedburner, which adds multiple features to your efforts to proliferate your content via blog. Check them out at Feedburner.com

Blogrolls

Located to the side of your blog is an area referred to as the "sidebar." The sidebar is an area for programs (widgets) that add additional information, hence value, to your blog above and beyond the content you produce. This is where the blogroll might go. A blogroll is a list of other blogs and their links that you recommend. You can populate your blogroll with anything you want, but mostly you'll want to put content relevant links in this area. You might also take some effort to contact other bloggers and ask to be put in their blogroll, kind of a reciprocal "rolling." Getting a link to your blog helps increase those relevance points.

Content Update Services; Pinging and Pingbacks

When you produce content, you want to have it found by search engines and directories. In order to have it found, it must be "submitted." Most of your content submission is best done by "pinging," one of several types of "content update services."

The word ping, when applied to search engine elevation, indicates a signal you use to communicate with search engines and directories analogous to an invitation. When you ping a search engine or directory, you are inviting them to crawl, or search through, your content, categorize it, and list it.

If you're not pinging, and you're not submitting links to social sites or bookmarking, your content is sitting stagnant on your blog or website, bringing minimal elevation value to your ranking in the search engines. To get out there and start improving your rank, you need to ping or submit, both methods of "content proliferation" that help you improve your ranking in search.

When you ping, you don't want to use the same directories too often;

most directories and search engines recognize abuse of this tool, and they can block you and your content or de-list it. When used carefully, pinging is an effective way to proliferate your content, and when abused, it can hurt you.

Material you want to ping should only be material relevant to the conversation you are having on your blog or website. Directories and search engines are crawling your content to organize it so people searching for the keywords and key phrases in your content can easily find it. Your content is part of a particular conversation and is best found when you stay within the boundaries of that conversation, so ping things that apply to your industry for best results.

I have heard it recommended not to submit content to directories and search engines more than three times a year. Pinging is all you need to do, so don't take the risk of getting de-listed or banned from certain sites. I know of one blogger who was banned from Technorati, the most popular blog index, because he submitted too often.

Many content management systems (free blogging software, such as Blogger, Wordpress, or Typepad) have pinging tools that make it easy to ping. Taking it to the next level, you can use some of these tools to ping as well:

- Ping-o-matic—twenty-two services
- PongPong ping—fourteen services
- Feed Shark—twenty-one services
- Pingoat—fifty-five services
- King Ping—twenty services
- Ping My Blog—eight services
- BlogBlip—fourteen services

WordPress, the free blogging software I prefer, automatically sends a ping to directories and search engines each time you create or update a post. In turn, Update services process the ping and update their proprietary indices with *your* update. Now people browsing sites like Technorati or Sphere can find your most recent posts!

Each time you post a blog, there is an opportunity for someone to comment on your post. If someone comments, you can accept the comment or trash it, and in any event, you can reply to it. A "pingback" is a notification your blog's content management system sends you when someone makes a comment (replies) to your post. A positive reply is an opportunity to establish a relationship and possibly get a link to that blog article or others.

If you ping it, they will come. Be sure to find out if your blog software offers automatic pinging—if not, consider changing or use one of the services above so your content can be found in directories and search engines automatically.

Trackbacks

A trackback is an alert you can set up that tells you, as a blogger, who has seen your blog post and referred to it in some other entry somewhere else on the web, either on their blog or someone else's, or within other content.

Content Creation: The Importance of Spreadable Media

A leading online trend is referred to as **Spreadable Media**. The term "spreadable media" was coined by Henry Jenkins, the Provost's Professor of Communication, Journalism, and Cinematic Arts at the University of Southern California, in his paper "If it Doesn't Spread, It's Dead". Spreadable media implies that for content to stay alive online - **which is to say being forwarded continually** - it must be shared within the social internet. Spreadable media has significant implications when it comes to business trends, from large media concerns all the way down to small businesses like yours.

So how does this apply to your business? Creating content tends to create greater visibility and awareness as it travels through unpredictable social media paths and is viewed by eyeballs of people who are in the market to patronize a business in your geographic vicinity. Your competitors' content is being disseminated widely and likely being consumed by people you would hope to draw into your business.

For the content creator, which may be you or your competitor or both, content is usually promotional in nature, but for the consumer of the goods and service you are promoting, the content has value and is viewed as a resource.

The consumer is likely to pass along a good resource because of the value they perceive their friends will receive from it, so the value in the content you create revolves around social interactions, and the social interactions occur based on the perceived value of the content you create. Content that doesn't have a perceived value, thus doesn't get circulated, has no business value. And businesses not creating content in this new media world are, in terms of drawing business from this huge social pool of eyes, dead.

The trend for people to share news after assessing its value, then passing it along to their social connections is getting stronger. This has significant and long-lasting implications when it comes to your marketing efforts. Take for example the people in your social networks circulating things to you, although sometimes not too selective, they are on the cutting edge of practicing this new, spreadable media. A practice that is sure to play a large role in the way media is spread within the global culture now and going forward. Is your business practicing spreadable media? If not, and you hope to attract the wave of people participating in the social internet, start writing!

CHAPTER 7:
SOCIAL MEDIA SUITES

LinkedIn

The following description of LinkedIn was taken directly from their homepage:

"LinkedIn is an interconnected network of experienced professionals from around the world, representing 150 industries and 200 countries. You can find, be introduced to, and collaborate with qualified professionals that you need to work with to accomplish your goals ... in a global connected economy, your success as a professional and your competitiveness as a company depends upon faster access to insight and resources you can trust."

When you join, you create a profile that summarizes your professional expertise and accomplishments. You can then form enduring connections by inviting trusted contacts to join LinkedIn and connect to you. Your network consists of your connections, your connections' connections, and the people they know, linking you to a vast number of qualified professionals and experts. Through your network you can:

- Manage the information that's publicly available about you as a professional
- Find and be introduced to potential clients, service providers, and subject experts who come recommended
- Create and collaborate on projects, gather data, share files, and solve problems
- Be found for business opportunities and find potential partners

- Gain new insights from discussions with likeminded professionals in private group settings
- Discover inside connections that can help you land jobs and close deals
- Post and distribute job listings to find the best talent for your company

I remember addressing a room full of eye-care professionals at a conference once. Had I wanted to meet any of them, it would have been easy to walk up and introduce myself. However, 99.9 percent of the time I am not in a room with three hundred other eye-care professionals. If I want to meet someone inside or outside of my industry, I could call that person and hope the call was returned, or contact him or her via e-mail and hope the e-mail doesn't end up in a junk mail folder or get deleted by accident—or perhaps on purpose. The ideal situation would be to have a mutual trusted friend or business associate who can make the connection. In regular life, these connections are hidden. As a member of LinkedIn, these connections are revealed. It's the old "six degrees of separation," but you can see each degree and who it is.

The following link takes you to an excellent short video that further describes the value of LinkedIn. Please visit http://www.youtube.com/watch?v=IzT3JVUGUzM.

LinkedIn is an excellent place to practice "searchial marketing. First, I encourage you to create your business "group" by going to the groups tab on the top, click "my groups," and then "create a group." Be sure to invite your LinkedIn connections to participate in your group and, to keep it interesting and growing, update it regularly with new discussions. Keep it fresh; think of it as a newsletter—let your connections know what's new and hot at your business.

An example of what will happen when your network is large enough is you will take an article you wrote, perhaps on your website or blog, and recreate a teaser in your LinkedIn profile, including a link back to the original article. People who are intrigued by your teaser will click the link and end up at your website or blog. It is these types of connections, these links, you want to increase, thus increasing your visibility in searches so customers are more likely to find you. While LinkedIn posts are nofollow, the searchial value of LinkedIn comes from your participation in the service provider directory. That is, you are raising your profile only within the LinkedIn community, which is a *very* large community of professionals who refer to one another.

LinkedIn Service Provider Directory

While I don't believe there's a ton of value using LinkedIn for business-to-consumer (B2C) marketing, I discovered B2C value in adding attributes and enhancing your listing under their service provider directory. Why? Because people who use LinkedIn *trust* the first and second tier connections in their networks and are likely to seek recommendations for service providers within these networks before looking elsewhere.

The service provider directory is an exclusive directory that includes service providers recommended only by LinkedIn user networks. Being listed in this directory immediately enhances one's credibility among other LinkedIn users.

To sign up for LinkedIn services, go to the LinkedIn services home screen and look on the right side, where you'll find pre-defined categories. Select the category that most narrowly defines your area of service. You will be prompted to make further choices by selecting the geographical location for the service you are looking for. There is a filter that enables you to select the degrees of your network in which you can access recommendations (near the top of the LinkedIn services page).

For example, the listing works like this: to find an optometrist in the Washington DC area who is recommended by your direct connections, you would:

- Select "doctor" in the Health and Medical category
- Click the "change location" button and enter the desired postal code
- Click the "1st degree" button to find recommendations from LinkedIn users you are directly connected to

Making a Recommendation

A way to enhance your chances of being found as a service provider on LinkedIn is to have people in your network make a recommendation for you. The more recommendations from LinkedIn connections, the higher you rise when people search for providers in LinkedIn. The LinkedIn service provider directory is a kind of internal search engine. Ask the people in your network to refer you by going to the LinkedIn services home page, clicking on the "make a recommendation" tab, and following the directions. Be careful to encourage them to select the appropriate category in which to recommend you, lest your recommendation get buried within some specialty where no one will find you. LinkedIn makes it easy to get recommendations. On the LinkedIn services page, click on the "request a recommendation" link on the top right of the title bar.

Choose the position within your LinkedIn profile that you hope to be recommended for, or use the "add a position" link to create the position and update your profile. Decide whom you will ask to recommend you—I recommend asking only those closest to you whom you can count on saying positive things about your services, and then create the message you want to send. Be up front and tell people in your message why you hope to have them recommend you and how important this is to your business. The more recommendations from people within your network, the more likely a second or third level connection will eventually find you through your mutual "trusted" connections and, hopefully, patronize your business.

Intellectually, LinkedIn is a major technological leap forward for society. Never in history were connections made visible via any other technology. LinkedIn brings tremendous value to any social network and is the most popular site that brings both business-to-business and business-to-consumer to the table in one network.

Facebook Introduction

One of the more useful suites for searchial media is Facebook. This wasn't always the case – at the time this book was written, Facebook did a major "about face" enabling much of it's content to be accessed by search engine bots. Facebook is implementing changes that are translating into searchial value for users, specifically through a recent deal with Bing. These changes create searchial value when people "like" your content or "check in" socially at your location. Facebook Notes, Page wall posts and other functions are now searched by search engine bots and provided searchial value. Facebook estimates that it has more than five hundred million users, 50 percent of whom log in on any given day. There are more than sixty million status updates every day and more than three billion photos uploaded to the site *each month!* It is a formidable network and is wisely leveraging it's audience to solidify itself as the leader in the social internet through partnerships with major search engine companies. Posting content on facebook will help your efforts to be part of the "searchial" internet and elevate your business profile. However, keep in mind at the time of this publication there is still a five-thousand-member cap on Facebook friends for regular accounts; groups, business pages, or "fan pages" allow unlimited numbers of friends.

Facebook defines their service as "a social utility that connects people with friends and others who work, study, and live around them. It helps you connect and share with the people in your life." Facebook is completely free, and although rumors circulate all the time that they plan on charging, those rumors are urban myths. You can share on Facebook by posting "status updates" or comments you leave on your profile page that announce information to your friends. You can upload photos and send a link to your friends or relatives to share. You can also use the photo function to organize your personal photo albums, keeping them secret. You can share web links, news stories, content, and blog posts with your friends. Facebook provides a helpful "event" creation widget that allows you to create an event invitation, distribute it to the friends you want to invite, and keep track of the RSVPs. Facebook allows you to have a page and/or group for your business, and you are able to market and organize events for your business through your group or page. Facebook has applications, such as iPhone apps, that allow you to do everything from play games to manage finances. You can organize your music and share it, use the weRead application to discuss books you have read, find others with similar interests, and the list goes on and on ...

When people reject the opportunity to participate in Facebook or other social media suites, it is often because they are concerned about privacy. These are usually people who don't do online banking for the same reason. Internet privacy is a legitimate concern for everyone and shouldn't be taken lightly, but personally, I feel a lot of the security concerns surrounding social media are overblown. When it comes to Facebook, you can *totally* determine what you do and don't want others to see. If there is someone out there, such as an old boyfriend or girlfriend, whom you want to remain invisible to, you can set your "preferences" so that a *specific* person can't find you (you can block the person, making your page invisible to him or her). If you want your posts to be available to a select group of your friends, you can set that as well. If you want to create an event and invite some friends but not others, you can easily adjust your settings to keep the event private, or to not show people who RSVP in the affirmative or negative. One of the best features of Facebook is your ability to control content as it appears to others.

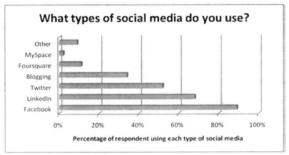

credit to Small Business Development Corporation of Western Australia
http://www.smallbusiness.wa.gov.au/surveys/#social-media

Facebook has become a useful tool for businesses. Most businesses create groups or pages they invite people to join. Your business Facebook page functions much like a blog—you regularly update content, post information about events your business is having, industry news, photos of the practice, etc. You may choose to use Facebook for business as a way of regularly updating patients, much like a newsletter or even as a recall system. You can send the same message to multiple people with just a mouse click, and the address list is not published, so privacy is maintained. You can create charitable groups or pages and use them to improve your business image. When you promote your Facebook page to your customers, you appeal to the younger generations; your image is boosted by appearing tech savvy and modern in the ways you choose to communicate. As many generation Yers and younger are communicating less and less by e-mail and more and more by social media, your efforts are welcoming and perceived as such by a younger client base.

There are many creative ways to utilize the tools Facebook provides for your business's benefit. People are more likely to participate in social media when a social good is attached to it. I had the idea that if I could benefit my small medical business and give something back to the public, perhaps I had a recipe for a successful Facebook page. My practice is Shady Grove Eye and Vision Care. I established a Facebook group called "Shady Grove Eye Care—Just Joining Provides Free Eye Exams to Those in Need." For every one hundred people who join my group, I provide a free eye exam to a person in need. I figured it wouldn't be difficult to give away a handful of eye exams at these numbers and if the effort snowballed,

I could stop it at any time. The group now has more than a thousand members, with more people joining each day. The only requirement of the people who receive the free eye exam is that they allow me to interview them on video and post their story on the page and on YouTube for everyone to see. By creating videos of their compelling stories of hardship, it creates more interest and provides continued growth of the effort. I post the videos on YouTube (only one posted so far—here is the link - http://www.youtube.com/watch?v=ZC47TGfUfUE) and post the link across my personal Facebook page, my business Facebook page, as well as promoting it on the "Shady Grove Eye Care—Just Joining" page itself, exposing my practice and my charitable effort to more than ten thousand people each month (one thousand members of the group, nine hundred of my personal Facebook network, nine thousand people who visit my blog each month, and I promote it to my 3,300 Twitter followers, all in the click of the mouse). Some people who appreciate the effort forward the link within their network, expanding the reach of my effort exponentially. This demonstrates the power of social media as a promotional tool, and the awesome power of the social media leaders facebook, twitter and youtube.

Facebook Business Utilization

Social media suites, such as Facebook, are useful for proliferating your message throughout your social network and can be used to filter viewers to the efforts you are promoting by incorporating links in the messages sent to your network.

Facebook for business involves getting your message out to a large audience. The larger your Facebook network (the more Facebook friends you have), the larger your broadcast audience. This is one reason why you should rarely reject a friend request. If the person requesting you to be a friend is someone you do not know or seems marginal in character for whatever reason, accept the request, but create a side-list and label it "acquaintances" or "strangers"—people you aren't familiar with—and stick that person in there. Within that list, you can filter which information these aquaintences have access to and which information they receive from you. If you are using Facebook for business, then you want your audience, or the people willing to read what you have to say in your posts, to be as numerous as possible, so it is in your best interest to accept everyone.

Remember, when someone accepts a friend request, he or she is giving you carte blanche to share information with him or her. But you want to be careful not to over-market to people or post spamlike content, or people will steer clear of you. Over time, if you are careful with what you post, and people enjoy your information, you gradually become a recognized expert on whatever topic you post and will draw people toward your efforts, which will in turn refer friends to you, and your efforts will have a quantifiable benefit.

Your effort might work like this:

- You write a post, let's say, on gastric bypass.
- You post it to your blog and create a short-link for your blog post by using a URL shortener like http://www.tinyurl.com.
- You update your Facebook page by writing a teaser in the empty status bar at the top of your Facebook page that says "What's on your mind?" (e.g., you would write *"Gastric Bypass—a new, safer method."*)
- Then immediately below the teaser, you copy the shortened URL to your blog post.
- Then you hit Share, and it blasts out to all your Facebook friends, informing them you have posted a new blog and giving them a direct link to read it.

You have now created a link from your Facebook to your blog and have immediately provided the link to tens, hundreds, or thousands of people (depending on the size of your network). Some of those people may have friends considering gastric bypass so they may e-mail the link to them, and someone may even reference the link on other social media sites. You have just added benefit to your blog *outside* of Facebook by promoting it within Facebook. The people who visit your blog need to leave Facebook to do so and may "subscribe" to your blog via an RSS feed, increasing the popularity of the blog in search, a point for your searchial efforts, and they may even be compelled to tweet the link to their Twitter followers—more searchial points for you. Search engines providing information to people searching for information using the terms "gastric" and/or "bypass" will likely have indexed your content under these keywords, increasing the likelihood your content will be found when people in your geographic region search this topic.

Additionally, to increase the number of people exposed to your post, you can directly request your friends "share" the link with their Facebook networks too by messaging them—if a post is interesting enough, it can become "viral" and gain a life of its own where it is broadcast from your post throughout many social media suites by thousands, potentially millions of people. The more your content is proliferated in the social internet, the more relevance and influence you appear to have to the search engines. The more relevance and influence you have as a producer of key word specific content, the more likely you will be found when people search for the products and services you offer.

Some content management systems (CMS), such as Wordpress, allow you to add a blog post, and when finished, click a button that publishes the post to your blog, your Facebook, and other social media networks at the same time! This diminishes the amount of work you have to put in to get your message out. I find this a helpful feature. And other sites like HootSuite allow you to manage multiple social media streams and blog posts in one place. These sites are called aggregators and are discussed later in this chapter.

The strategy to using a Facebook page for business is to disseminate fresh and interesting content to your connections. Discussions on Facebook happen around content, usually original content. Followers want to read original content on interesting subject matter and don't want to be spammed with information on the products or services you offer. That doesn't mean you can't promote, just do it sparingly. My advice is to promote your business or products less than 20 percent of the time. If people like what you're saying on Facebook, they are more likely to tolerate promotion from you if promotions are few and far between. Facebook alone is not enough to wage an effective social media campaign—you need original content and other tools to interact with Facebook to get the most out of it. As your reputation as a thought leader grows, you might even find yourself interacting with industry contacts who want to learn from you or use your content, another opportunity to get link juice by asking them, in return for using your content, to create a link to your blog or website, not to mention other opportunities that may arise from a relationship with a corporation in your industry.

Facebook offers you the opportunity to create a page for your business, and this gives you the ability to separate your "social" (personal) Facebook group (friends and acquaintances) from people you want to market to. I suggest you set up a business page and delegate building the page's followers to your staff. Whenever a new client enters your business, ask if he or she is on Facebook, and invite him or her to join. You can send business news with the click of a button instead of the cost of printing and using postage, and a Facebook message is received much more eagerly than an unsolicited piece of mail. When fans make comments, be sure to respond to them, or "like" them by hitting the like widget. This can be a time-consuming process, but this is the interaction you or a delegated staffer need to be involved in because Facebook is a social network, thus only effective when you are social within it. It is a good idea to have pages managed by different members of your staff; different input creates richer, more varied content. Different staffers have different communication styles and bring different value to the table in terms of the cumulative effort for your business. Not all staffers are going to be capable or even talented writers, so be selective with who you delegate to. I have certain members of my staff create content, but I maintain final edit approval and don't enable them the ability to post without running it past me first.

Images and photos are important to your Facebook effort. Facebook has a function that allows you to tag photos; that is, to put a label on them. If you want to grab someone's attention on Facebook, post a photo and tag it with his or her name, whether her or she is in your network or not! A notification that he or she has been tagged will be sent, and it usually will result in him or her visiting the picture to see what's up. You can use the tagging function creatively to drive business. Justin Bazan, an optometrist and owner of Brooklyn's Park Slope Eye is an innovator in creative tagging and uses the function to create contests, giveaways, and generally raise awareness of his Facebook efforts. His use of this and other Facebook functions is highly creative, and I strongly recommend you monitor his efforts to learn from them.

Remember, creating a link to your content and posting it on your Facebook page can drive visitors to your website or blog from Facebook, so it has business value, but as Facebook is nofollow, it will not result in a bump in search engine ranking. Thus, Facebook efforts do not qualify as having traditional searchial value for your Internet marketing efforts.

Facebook Places

Facebook offers an incredibly useful free business listing called **Facebook Places**. The service enables you to advertise a coupon for free for unlimited impressions of your ad, and if you choose, enable viewing on mobile devices so your coupon shows in searches that occur through facebooks mobile website. Facebook Places is growing in importance as the mobile web becomes more popular. Facebook gives you the opportunity to describe your products and services in great detail and for free. Don't limit yourself by leaving out attributes of your business. Tell potential new clients what sets you apart, don't be shy. Be sure to include links to your **blog** and your **Twitter** account.

Facebook Places can now be merged with your Facebook business page, and I recommend doing this as it consolidates your facebook marketing efforts. Be sure to use tags to effectively market your facebook places page and promote it throughout your brick-and-mortar business.

Facebook Markup Language

Facebook Markup Language (FBML) is a way to create applications compatible with Facebook via an application that interfaces with Facebook. For example, a Facebook "API" (application interface—a program that runs separate from but accesses information within Facebook and can have a "tab" displayed on Facebook) can be designed to add almost anything you can dream up to the Facebook network. Many of the games and quizzes people send you are applications created through an API. These applications can be used for personal use, business use, or in any other way people can come up with to enhance the Facebook experience. You can post these applications on the user's profile, Facebook "canvas," and within the news feed and mini feed.

If you have a creative idea for a Facebook "add on," I recommend finding a developer you can describe it to and pay to have it developed. You can go to developers.Facebook.com for resources or try eLance.com if you want to outsource the development overseas and save some money. Use the application to promote your effort, collect lists of people to market to on Facebook, or drive other marketing efforts. I would be happy to refer you to someone if you don't know where to turn.

Facebook Advertising

Facebook has paid advertising that is effective and affordable. As of this writing, Facebook ads cost less than Google Adwords ads, although Facebook ads are limited to the Facebook network, while the Google Adwords Ads go Internet-wide. Basically, your ad appears on the right side of people's pages. Facebook offers a fee structure of either CPC or

CPM (see Chapter 1 for more information on how to use paid advertising on Facebook).

Twitter

Twitter launched as the Internet's first "microblogging" site. A microblog is a blog post whose content is condensed to one hundred characters or fewer. "Tweeting," as it is called, is the act of posting on the social media suite Twitter.

On this information highway, where you get run over if you are too slow to absorb information and where attention deficit is an asset, it is helpful to receive information in the form of one-liner "microblogs," detailing information you are seeking, and enabling you to delve further if you wish. Or move on to the next bit of information if you don't. When you look at all the information that comes at you through social media, the value tweeting brings to the table quickly becomes apparent. It is an effective way to sort information. A tweet can contain links to blogs, videos, photos, or almost any linkable content.

For my searchial marketing efforts, the value of Twitter is twofold:

1. It acts as a filter, allowing me to quickly and efficiently scour through the information my sources post without having to delve to deeply and spend too much time, and
2. It allows me to get my message out in succinct, readable form and draw people toward my searchial media efforts, while active tweeting keeps me in Google bot sights, adding relevance to the content I link to. Twitter is dofollow—what you tweet is picked up by the bots.

I'll admit, it took a while to understand how to effectively use Twitter and even longer to appreciate its value, but now it is a cornerstone of my searchial media campaign. It also helps that I am the poster child for attention deficit disorder and enjoy getting pelted with thousands of messages.

The idea behind Twitter for business is to attract as many connections as you can. When people connect with you, they are giving you carte blanche to push information into their lives. Like the other social media suites, your best strategy is to promote helpful information that drives people to your blog or your other social media efforts. *Don't annoy your followers by trying to sell things to people or post less-than-credible content. These are sure ways to scare potential followers away or lose current connections.* Like Facebook and LinkedIn, the more connections you have on Twitter, the larger the group you can get your message out to. When you tweet, people may read your message. Some of those people will "connect" with you. An even smaller percentage of people will find your message interesting enough to click through to your blog, your Facebook page, or your LinkedIn page to see what else you have to say and may connect with you there also, increasing your connections across the board. An even smaller group of people may find your tweet so compelling they'll re-tweet it to their network, and people on a totally separate network than yours will be exposed to what you have to say and may end up connecting with you as well. An interesting or compelling post can result in hundreds, thousands, or, in the case of actor Ashton Kutcher, even millions of Twitter connections. If your tweet is of interest to another person, a link within the tweet allows them to travel to get that information. If that link is to one of your social media efforts, you get search engine credits for each time someone visits your efforts.

Like Facebook, there will be a small group of people who might find you and live close enough geographically and show up at your business, but gearing your efforts toward this market to generate new business is not efficient. Use Twitter to connect with your customer base. The grassroots effort I described in earlier chapters happens often within Twitter. Customers talk on Twitter, and businesses need to be listening and interacting. Getting in front of your clients on Twitter helps keep you in their minds. Businesses like mine, where people visit typically once a year, might learn about an eyewear sale through my blog or tweets and be compelled to increase their visits per annum to my business. To generate *new* business, your goal on Twitter should be to create content that is recognized by the bots, causing you to float to the top of the organic search results when someone in your region searches for a "doctor in Seattle Washington," or wherever your business is. Think *of Twitter as medicine, and your tweets as pills. Twitter makes your blog or website or other "linked-to" content "healthier"—i.e., more visited by people searching for it, like medicine, and each tweet is a "pill" that delivers the "cure", gaining visitors, links, and ultimately relevance to float you to the top of the organic listings.* You will show up ahead of your competitors in the major search engines because you have social media campaigns around your specialty that are up-to-date and relevant to the subject matter your clients are seeking.

When you tweet, your tweets are rapidly picked up by search engines and placed higher according to the keywords, key phrases, and links inside your tweet, and they rapidly are replaced over time with more relevant tweets. Thus, when practicing searchial marketing, it is imperative to keep a steady stream of tweets with relevant keywords and phrases out there to keep yourself on top of search results.

A second goal might be to interact with followers who are in your geographical vicinity. This is following the grassroots social-media strategy I wrote about earlier in the book, and is used mostly to build existing customer loyalty and provide channels that make you appear relevant, current, accessible, and trustworthy. This encourages your customers to frequently interact with you, building trust, which strengthens the business relationship. You can collect Twitter followers like you collect patient addresses and e-mails, notifying them of promotions and events.

Tweeting Tips

- Your tweet shouldn't use up the entire 140 characters—you should leave approximately fifteen to twenty character spaces so people can re-tweet your content, or else your tweet will only go out to your followers and won't have a chance of becoming widely disseminated.
- You need to *get involved* on Twitter and stay active—stagnant Twitter handles will dwindle in members. If you are inactive for as short a time as a week, you will see the number of followers decline, so stay active!
- Don't push products or services too much on Twitter. It is okay to hit your network with a pitch every once in a while, but people selling on Twitter appear cheesy, and if you are perceived in this manner, it will hurt your campaign.
- Post links to your blog with teasers like *"New diabetes therapy shows promise ..."* and attach a shortened URL (obtained by visiting http://tinyurl.com) to your post. People interested may end up at your blog and may subscribe to it, increasing the blog's popularity, or re-tweet your blog URL so your blog's popularity increases, increasing its chances of being found in search engines. Don't give away the farm in your tweet—keep it short and simple and use it as a tool to draw people deeper into your social media effort.
- Participate in the discussion.—Getting re-tweeted is nice, but re-tweeting someone else's post is also nice. When you participate

in the discussion through re-tweets and replies, the major search engines know and "prop" you up. When you don't participate, you are nowhere and get no search engine credit. It is helpful to reply to people and re-tweet them; you get respect and develop interesting relationships that open doors and help your search engine rankings in the process. Don't use the private reply option much; it doesn't add to your viral efforts, as it is only seen by you and those you message, like, and e-mail.

- Set up a nice profile.—Always use your real name so people know whom they are dealing with. Put a colorful and interesting background on your home page and have a powerful introduction message.
- Follow people and they will follow you.—It's Twitter courtesy and an easy and quick way to build your Twitter network. Don't use the programs that cost a lot of money and promise you followers, you just end up pissing people off, getting a lot of junk in your Twitter account, and having to trim the list eventually anyway.
- Keep your follow-to-followers ratio even, or have slightly fewer people you follow than follow you. This gives your account the appearance of being healthy, and people are more likely to follow you if you have a healthy account than if you appear spammy.
- Get creative. I struggled for a long time to gain followers, and as I became more familiar with Twitter, I learned how to use its tools, such as hashtags (#), to increase the popularity of my efforts (see section on *hashtags* in Chapter 5). The first hashtag I created was #badvisiondecision. Whenever I would encounter a patient who made a decision that could negatively impact his or her vision or eyes, I shared it on Twitter (of course not mentioning names).

For example, I posted the following information after I had a patient who got hot sauce in her eye, and I tagged it like this: *@EyeInfo Hot Sauce on Finger-In-Eye #badvisiondecision*

Dr. Nathan Bonilla-Warford posted the fact that smoking was bad for people's eyes like this:

@BrightEyesTampa smoking #badvisiondecision

I also created #MedicalMonday for people to acknowledge their healthcare providers. On Mondays, people type the Twitter handle of their favorite healthcare providers or of people who are posting good health information on Twitter and follow it with a #MedicalMonday hashtag, where the tweets are grouped together for the benefit of everyone who wants to review them. #MedicalMonday has helped me introduce myself to hundreds of people tweeting in the healthcare area, and I have established many contacts and received thousands of hits on my blog through this effort. I have lately taken a break from the #medicalmonday efforts due to the time involved but will resume it at a later date.

These efforts, among others, help boost my Twitter profile and keep me active as a member in the Twitter community, even when I can't actively post. People appreciate the information, and a certain percentage of them, when they come across your information, will appreciate it enough to follow you, and that is what you hope for.

Selecting a Username on Twitter

Your username shouldn't be random; it should be simple, descriptive, and memorable. Try to pick a name that is short, as longer usernames invite typing errors that may cause someone to miss you. Numbers in the name aren't recommended for the same reason. The best name is your own, or a name associated with you or a business you are promoting. Use your real, full name in the description on top; people want to know you're a real person and identify with you so they can converse. Be sure you pick the Twitter username you plan on sticking with—changing it can confuse followers and cause drop-off, plus, you can't access your old timeline anymore. If you really want to use a different username, you should create a second account and manage both—easy-to-use Gmail is the only e-mail service that allows you to manage multiple Twitter accounts. *Do not* use underscores in your twitter handle; underscores aren't recognized by the Google bots, and this might negatively affect your Twitter handle showing up in searches, taking away from searchial benefit.

Sidebar

Your sidebar is the area on both sides of your Twitter page. Think of this area as area place to market your efforts; post links that advertise your blog, website, or other social media sites. Right now, you can't put click-through links anywhere but in the description on top, but that doesn't mean you can't display links you want people to hand-type into their browser. I recommend hiring a professional to design your sidebar backdrop images and help you place content in the sidebars, as it is a very important part of your efforts. Be sure your contact information is easily visible in the sidebar.

Choose to obtain an e-mail from a new follower and when you get a direct message, so you can attend to these followers and acknowledge them. Direct messages are kind of out, and don't have a very important purpose on Twitter, so you want to phase them out of your plan eventually.

You can now receive tweets as SMS messages. To set this up, press the devices tab under settings, enter your phone number, and click the checkbox that allows you to receive tweets on your phone as SMS messages. You'll then receive code 40404 in the US; 21212 in Canada; 49 17 68888 50505 in Germany; 46 737 494222 in Sweden; 5566511 in India; and +44 7624 801423 everywhere else, and your phone will be set up to send and receive tweets.

Experts on Twitter

Experts and highly followed tweeters are important. Experts are great connections on Twitter, and finding the experts you need is a great benefit. Many people tweet about topics that are of use to other people, and the experts you are interested in are great targets to connect with, as they are easy to engage in conversation with and likely to follow you back. Find experts by searching Twitter using search.Twitter.com. The Twitter search bot integrated into Twitter can be used too; it isn't very strong or efficient, but using it is helpful in learning to navigate your way around Twitter. Instead of following everyone on a particular topic, pick the key people and follow them—it will expose you to the others, and as you converse with the top experts, many will gradually see your name in similar conversation and may follow you back as a result; this is a common strategy to gain Twitter followers, a usually successful strategy and a legitimate way to build a great following of high-level Twitter connections. Reading tweets from experts is also a great way to learn about social media and get ideas for content you want to create.

Thought Leadership on Twitter

It will enhance your efforts to make yourself stand out and be seen as an expert on Twitter. Experts gain followers and, as we have established, followers are good for many reasons. Being involved in a conversation on a particular topic that many people can see starts to make you look like an expert, and people will pick up on this and start to recognize you, at least as being part of a particular conversation, and at least as someone who is an authority on subjects in particular conversation lines.

Use the direct message function sparingly. First, direct messages can't be viewed by the Twitter populace, so your messages won't bring any value to the broad conversation. Second, direct message boxes typically contain too much spam and are generally ignored. Don't engage your followers too frequently—this is the cause of drop off. Know when to contact them—when you have good information, something of value, or something interesting or funny to say. According to Joel Comm, in his book *Twitter Power 2.0*, "Firms that get social media wrong look like interlopers, uninvited guests who have gate crashed the cool people's party." Not "getting it" can have a negative impact on your Twitter marketing.

Building Your Following

Use Twitter tools to find the people you already know on Twitter—use the "find people" link at the top of the page, enter the information, and Twitter will do the searching. This only works for web-based e-mail like Gmail, AOL mail, and Yahoo! but not for private addresses.

Comm, a marketing expert and author of *Twitter Power 2.0*, recommends searching Outlook by opening a free web-based Gmail account and exporting your contact list from Outlook and importing it into your new account, and Twitter can automatically search these. The number of followers you get depends on how big your contact list is.

As for promoting, you want to put your Twitter handle in all your e-mail signatures and on your business card, put a sign at the reception desk so people can tweet you—run a contest. Jason Cormier (@jasoncormier) runs a social-media agency and helps you organize contests on Twitter.

Twitter Tips

Follow tweets and observe the types of conversations, etiquette, and symbols like hashtags. Observe how links are used before you start requesting friends. Observe the things other Twitter users are likely to respond to and what they are not, to make your conversation more efficient in terms of growing. What you want to have happen is to have a conversation with someone (Mr. X) who has a large following, because your Twitter handle will be observed by many of them. It is likely some of them will ask the question "who is that tweeter who is so important that Mr. 'X' is corresponding with them?—I should be following that person." Use your sidebars to market—they are a very powerful tool!

Twitter is more social than Facebook because you don't have to be friended or accepted before getting involved in the conversation—resulting in much more open conversation with millions of people. You also don't have to add someone to your following list to communicate with him or her. It is also more searchial, because what you say gets indexed by the major search engines, boosting your position in search.

Twitter and Your Blog

Twitterfeed.com, Bit.ly, and SocialOomph are services that link to your blog and automatically update your Twitter feed with your latest blog headlines—when using Twitterfeed or other proliferators, be sure to craft your blog headline within the maximum 140 characters so the whole headline gets into the tweet, and it's probably a good idea to make it fewer than 126 characters, so people can re-tweet with their handle attached and still get the whole message across. To learn more about services I use and depend on, such as socialoomph, visit http://wp.me/pxAmm-zu and click on the images.

Using Your Followers for Ideas

Twitter is a great place to learn. People inside and outside of your industry are always sharing links, thoughts, quotes, music, videos, and a host of other media. Find connections in your business space and watch what they post. When a particular tweeter or connection interests you, try to mine them as a resource for blog ideas, and don't be afraid to ask them to guest post. You might find a great linking source. You can also post requests to see topics your followers would like blog posts on; don't make your requests boring or dull; try to make them funny and interesting, as this will increase the likelihood of people responding to them. The best tweeters are those who are already talking about you. The "track KEYWORD" tool can be used to find these people—type track KEYWORD into your IM or mobile device, and every time someone types in your @username in a tweet, you'll get a message to that effect. http://www.monitter.com is another way to follow conversations about you.

Twitter Analytics

Twitter statistics hosted by Twitter only tell you how many people you follow, how many people follow you, and how many updates you posted. You can create your own analytics by posting different kinds of tweets (e.g., funny tweets, tweets with information, tweets with news headlines, etc.) and observe and track the response—the type that gets the most re-tweets you can use as a tool when you really want to make an impact, but this is a lot of work. A better way to get statistics on Twitter is to use www.trendistic.com—it lets you find more followers by tweeting about popular topics—the more current and "hot" the topic in your tweet, the more likely people will find it interesting, click on it, and hopefully follow you. Tweetbeep.com sends out alerts when people are talking about the keywords you are interested in. Twittercounter is a site that lets you track metrics like the results of your tweets (how many re-tweets each tweet brings in) and which tweets of yours continue to generate interest. TwitThis is a content proliferator plug-in that allows you to send the URL of your blog to your Twitter feed automatically.

Searching Twitter

The Twitter search tool is confusing and difficult to use. You can find information you are looking for easier with these Twitter search tools:

Search.Twitter.com—Probably the best tool to search past tweets, specific people, terms, or phrases. Very easy to use; just go to the site http://search.Twitter.com, enter your search, and voila!

Monitter.com—Monitter is a widget you place on your website or blog that lets you monitor all the posts on Twitter for specific keywords, names, or phrases that pop up. You can do the same thing with hashtags on Twitter. It's a really fun tool to play with and great for spying on the competition, following what people are saying about you or your business, or following topics of interest.

Paid Advertising on Twitter

A relatively new phenomenon a number of services are now offering is to deliver ads to Twitter users. I have lately seen creative advertising sales from people offering to send your message to thousands of people for $5. Some advertisers automatically generate their ads within your timeline, taking some control away from your Twitter feed efforts. SponsoredTweets.com is one of the more popular of these services. SponsoredTweets.com lets viewers know that the tweets they are reading are ads, so it does not appear they are trying to trick their followers. You can sign up for sponsored tweets by going to http://spn.tw/r2L4i . There are also tweeters with significant followings who will message their followers with your advertising for a fee.

Third Party Tools on Twitter

SocialOomph

One of the strategies used to gain followers on Twitter is to follow the people who follow you. It is Twitter courtesy, and it demonstrates interest in the people who demonstrate interest in you, a very social exercise. This can get to be tedious if you start to get enough followers, so sites like socialoomph.com can help. With SocialOomph.com, you can set up automatic follows, so when someone follows you, you end up following him or her without remembering to do it manually. Another great feature of SocialOomph is that it lets you set up tweets automatically—have a favorite post you want to proliferate periodically? No problem! SocialOomph is the tool for that. It takes some pressure off the need to keep your feed busy and lets people know you're participating in the conversation, even though you aren't (really)! To learn more about socialoomph visit http://wp.me/pxAmm-zu and click on the link.

Social Bookmarking and Tagging

A bookmark, as everyone knows, is an implement placed between the pages of a book for the purposes of holding your place to re-reference, share, or return to it later. Sometimes it even has a piece of yarn attached to it or a Hallmark quote on it.

A *social* bookmark is a bookmark on a website (as opposed to a browser) in the form of a link to a webpage one wants to re-reference or share. *Social bookmarking* is an important part of any searchial marketing campaign for small business.

Most of you are probably familiar with the bookmark or "add to favorites" tab in your web browser. This system, while helpful, quickly can become difficult to search, and the bookmarks are only accessible by one computer (the one they are saved on). The advantage of social bookmarks is that saved content is done so by humans instead of search bots and spiders, on a website accessible by everyone on the Internet, through tags. Think of social bookmarking as a card-catalog on steroids.

You can use social bookmarking to get your web pages or blog into search engines under tags people might search for your content, and you can use it to organize and follow websites and other resources you hope to revisit. Bookmarks are used by people to choose and save the best pages to view later. The sites not only help proliferate the content you want to share, they organize content you want to keep track of. View a great video to introduce you to the concept of social bookmarking here: (http://www.commoncraft.com/bookmarking-plain-english)

How to Gain Searchial Benefit from Social Bookmarking Sites

Some bookmarking sites are dofollow and provide searchial marketing benefits, and some are nofollow and do not. However, the significant searchial benefit of all social bookmarking sites is the ability to write and place your own tags, the keywords and key phrases you believe people use to find the products and services you offer and use in your content in the hopes they will find you. The proper tags help search engines pull relevant information from keyword and key phrase searches more efficiently. Results show up bookmarked on social bookmarking sites faster than search bots can index them, so they appear in the bookmarking sites before they do on Google or other search engines in general. The most popular bookmarked pages provide relevance value for the site they came from; the content that is heavily bookmarked through several social bookmarking sites is seen as more of an authority on the content than one that is not. Also, bookmarking is a great way to index pages that are deep within your site, making those pages more navigable through search and the tags people use help search engine algorithms classify sites more efficiently.

Social bookmarking sites that do not use nofollow (see section on nofollow). The *major* social bookmarking sites, such as Stumbleupon, and Ma.gnolia, use nofollow attributes, but there are other search engine elevation benefits to using them, such as external meta data and lots of traffic, so you should still consider bookmarking them. Here is a list of bookmarking sites with no nofollow attributes at the time of this writing:

- Digg.com
- Technorati Faves
- Listible
- Slashdot
- Flickr
- Furl
- Propeller
- Searches
- Yahoo! My Web 2.0

Tags are descriptors that one can add to one's bookmark that describes the content they are bookmarking. For instance, let's say you post an article entitled "New Therapy for Macular Degeneration," where the text reads "ophthalmologists develop an anti-VEGF molecule that may halt the worsening of wet AMD..." on your blog and bookmark it using a social bookmarking site. You will be provided an opportunity to tag your bookmark by entering phrases that relate to your article. Tagging and descriptions used in bookmarking is referred to as "external meta data" and has great value in search engine elevation. Your tags may include:

- *ophthalmologists*
- *anti-VEGF*
- *wet*
- *AMD*
- *macular*
- *degeneration*

When you want to search your bookmarks, you can search easily on the social bookmarking site via the tags you applied. Your search will also bring up relevant content bookmarked with the same tags by others, providing you access relevant content that you might have not found or would have had difficulty finding. You can't do this in your bookmarks or add-to-favorites tab on your browser. More importantly, your bookmarks are public, so when someone out there wants to find information on, for instance, anti-VEGF pharmaceuticals and wet AMD, and uses the tags you wrote as descriptors in his or her search, pages bookmarked with these terms are likely to show up together and ahead of pages without these tags. References tagged with similar terms and greater frequently by more social bookmarkers are also picked up by browsers as being more relevant, ranking them higher in web search. It is important to know that the web spiders and bots that search the Internet do so only periodically and don't always pick up relevant pages, so by tagging your own bookmark, you are increasing your odds of having your article show up in the search engine, and sooner! Now that's searchial! One of the drawbacks of tags includes the misuse of tags, misspelling of tags, and susceptibility of social bookmarking for use by spammers. Tags are important to your efforts to be found in search as a way to categorize you, so people searching the content you hope to be found within can find you.

How Tags can Affect Your Business

People use tags to classify things, so people will tag your content the way they see fit. You cannot control what people say about you online, and you cannot control which tags they apply to you. People used to set up websites with nasty URLs to get their point across; I've seen the URL USAirsucks.com and have even seen this strategy applied to individuals and small businesses; now an unhappy patient can tag your content with the key phrase "bad customer service" or "poor judgment." Just think about the negative impact this can have on your business. Tags can be used to undermine your efforts, so pay attention to them. It is okay to tag your own content to help people searching for you or your services to make finding you easier.

Social bookmarking sites that are popular include delicious.com, stumbleupon.com, reddit.com, faves.com, and diigo.com. I recommend you explore social bookmarking by signing up at these sites to start and bookmark at least three or four of the pages of your website or blog that you feel are the most important (i.e., want to drive the most traffic to). Be sure to tag them as well! In Chapter 6 we discussed the nofollow attribute; be sure not to spend too much of your time bookmarking content on sites that nofollow, except delicious and stumbleupon.

Bookmarked content increases in relevance based on the number of times the content has been bookmarked by other users. More and more users are using social bookmarking as a tool to make their website more visible.

Social Network Aggregation and other Combination Tools

The more social networking sites you sign up for, the harder it is to keep track of everything, and at a certain point it becomes impossible. How do you reconcile the fact that it benefits your online efforts to sign up for as many sites as possible, yet the more sites you participate in, the more confusing and difficult your online marketing efforts become? Time for you to "Lifestream."

Social network aggregation, also known as "Lifestreaming," occurs when content from multiple social media streams (Facebook, Twitter, etc) is "steered" or "fed" onto one page for the purpose of organizing content to a single location for easier access. Social media sites, called aggregators, exist for this purpose. Different types of aggregators that stream data across different types of networks exist, allowing users to aggregate news feeds, microblogging messages, friend updates, bookmark organization, etc. This is referred to as Lifestreaming, as people use it to manage their online life more effectively.

The concept for social network aggregators originated out of a parallel challenge in the instant message space. Trillian, developed to connect multiple messaging networks (such as Windows Live Messenger, Skype, etc.), launched in 2000. The latest free version is called Trillian Astra, and they recently released the first mobile application to iPhone. Trillian was the model for social media aggregators.

FriendFeed

One of the first and the most successful aggregators, FriendFeed, can consolidate more than fifty-eight services and counting. FriendFeed allows you to pull all of your social output into one central location, which provides your friends a single place they can keep track of you, as well as for you to keep track of your social media feeds. Friends can subscribe to your FriendFeed to see all your updates wherever you've posted them. This makes it a great place to keep up with friends but also a handy tool to spy or act voyeuristically.

Advantages: There is no need to refresh your browser with FriendFeed. Updates stream in automatically.

Disadvantages: You can't include friends or contacts unless they use FriendFeed also.

Seesmic

Also known as "the Twitter' of video," Seesmic can integrate multiple

Twitter accounts and Facebook. It also offers web-based tools like URL shorteners and image-sharing services. Seesmic can be used by different mobile devices or as a desktop application and is now becoming more of a social-media site itself, enabling you to build a community around it. Seesmic and Tweetdeck offer similar services (read about Tweetdeck below).

Advantages: Twitter user interface easy to use. Really geared toward video and Twitter.

Streamy

Streamy is a social media site aggregator that also feeds instant messages and blogs to one dashboard. With Streamy's Status Update Tool, you can post your updates directly to Facebook, Twitter, and many other social media sites with the touch of the mouse. The updates for your services are housed within tabs and don't all appear on one page. Streamy allows you to customize each tab, enabling you to promote your preferences within each tab and demote the information coming from the particular social media stream that is less interesting or important to you; for instance, think of the Facebook tab as your own personal Facebook page that you can modify any way you want.

As such, Streamy doesn't merely act as a dashboard for all of your social media sites; it also allows you to create a custom dashboard for every site, offering a heads-up view of what's going on at each individual site based on what's important to you.

Streamy has a "What are you doing?" status box visible on the right side of the page that allows you to write a post and simultaneously broadcast it to all the sites you follow in your Streamy feed.

Advantages: Interesting way to keep track of your social media efforts.

Disadvantages: Not truly aggregating things in one easy-to-view page.

Flock

Most social media aggregators are web services that you sign up for. Flock is actually a web browser that sits on your desktop. Think of it as a Firefox, Bing, or other browser. The difference between Flock and your typical web browser is that Flock is a social web browser designed specifically to integrate the most popular social media sites. Flock features a "people" sidebar. Every time you log on to a social media site on a service supported by Flock, your site appears on the people sidebar, and you can click the icons to make them appear in the browser. It also gives you the option to click "all" and view everything in a chronological feed sequence. You can post simultaneously to multiple supported sites as well.

Advantages: Using a web browser to aggregate social media is easier because most social media usage takes place in a web browser. A huge advantage is that you don't have to provide your login information to third party social media sites, which helps preserve your online security.

Disadvantages: To use effectively, you really have to shift all your web browsing to the browser and discontinue using the browsers you currently are used to. You can also only use Flock on computers where you can download and install the application, meaning you might not be able to use it at work.

HootSuite

Another web-based (browserlike) tool, HootSuite is one of the most popular aggregators. It is convenient to log into from any computer or mobile device. Currently, HootSuite supports the major players in social media, such as Facebook, Twitter, and others. It also supports Foursquare, a geo-targeted mobile application that is greatly increasing in popularity.

In HootSuite, you can:

- Choose between live updates, or pre-schedule posts and shares in advance.
- Add custom link parameters for tracking clicks and gathering information on your audience.
- Upload images, video, and files right into your messages.
- Connect to your RSS and send your blog to your social media streams.
- Use the HootSuite Hootlet from your browser toolbar to share pages and information quickly.
- Keep up with HootSuite from your iPhone using handheld integration.

- Create and customize columns that can be dragged and dropped in any order to your liking.
- Harness the clutter and organize your social streams into news, keywords, friends, and more.
- Grab code from HootSuite to embed search columns directly to your website.

TweetDeck

TweetDeck, similar in many respects to HootSuite, has the added advantage of being particularly customizable. The systems that support HootSuite are diverse and include Windows-, Apple-, and Linux-based systems.

TweetDeck features include:

- Choosing how you'd like your interface to appear and making it your own.
- Setting up your columns to show you only what you want to know.
- Getting alerts for new tweets, mentions, and direct messages. (This is an excellent communication management feature.)
- Deciding whom you should follow, unfollow, or not follow, as well as reporting spam and marking your favorites.
- Incorporating the bit.ly auto-shortening URL for tweets and image uploads.
- Setting up TweetDeck to suit your personal tastes, and keeping it that way regardless of whether you access it through your laptop or your handheld.
- Tracking your favorites and organizing them into Twitter lists right from your dashboard.
- Seeing what's hot with local trends and Twitscoop.
- Speeding up your actions using keyboard shortcuts so you can maximize your time with TweetDeck.

Cliqset.com describes themselves as "a website that lets you manage multiple identity streams in one interface, making sharing and discovering information more manageable." Cliqset is one of the few aggregators that enable aggregating placestreaming sites like FourSquare. Placestreaming refers to the stream of content that emanates from a place with a specific geographic location. Cliqset.com claims to be able to integrate with more than seventy different social services, including Facebook, Twitter, LinkedIn, and Foursquare.

Socialite—Particularly useful for Mac users, the Socialite homepage has a Mac-like design and makes it easy to manage your social media accounts from your Mac. This is the only aggregator mentioned that you have to pay for ($20 per license). It doesn't sort things into windows or columns, and the face is uncluttered and easy to navigate. You can use Google Reader or other RSS feeds and popular social networking sites with Socialite.

The Aggregator "Aggregator"

There is an industry joke that makes fun of the fact that there are so many social network aggregators, they need their own aggregator. I know you are probably falling off your chairs laughing, but Google is actually working on this with a project called SocialStream. Google is not alone in the race to deploy the first aggregated social network. Blue Swarm and Wink have some features to aggregate social aggregators. Other networks are competing for this Internet "holy grail," including Mozilla, but there are no standouts at the time of this writing. One thing is for certain; people using social media tools will jump ship to use social media tools that are friendlier, so should one of these "open source" social media aggregator aggregators get some traction, look for the entire face of social media to change virtually overnight. The company that figures this out will become an incredible powerhouse based on its ability to data-mine users' personal information and likely would become an Internet giant. But I digress ...

Microsoft has recently launched a social-media aggregation tool it calls Spindex, in a closed beta-version. In addition to handling typical feeds from Facebook, Twitter, etc., such as the feed dashboards discussed earlier, Spindex will handle RSS feeds, bookmarking, and its own search engine—Bing—and likely will add many more services.

Mashups

Mashups are sites that combine a number of web 2.0 tools or several sources to create one data set or program. Mashups are supposed to offer value by bringing like content together easily for easier surfing and utilization purposes. There is a trend online to produce web applications using a format called application programming interface (API). An API enables a software program to interact with other software. Basically, it is an interface between two software programs that makes them compatible with one another. Wikipedia makes the helpful analogy that an API facilitates interaction between different software programs similar to the way the user interface facilitates interaction between humans and computers. Mashup tools play a major role in the evolution of web 2.0 and social software. Data can be "mashed" together to create a new and distinctive source of data. Business mashups occur when two businesses combine resources via the API and allow for collaboration. Twitter is a common platform that integrates API, as they are open and accept almost any API to interface with them.

Schedgehog.com—Finding Lost Revenue

One of the primary revenue drains in any practice is vacant appointment times. According to the Medical Group Management Association, physicians with busy practices lose 12 percent of available appointment times daily, due to patients that don't show up, who reschedule a few hours ahead or cancel at the last minute. This is a significant drain on income; based on this figure a practice with one doctor can lose $32,000 or more per year from missed, cancelled or rescheduled appointments. The problem is these openings are usually same-day, and coordinating people into newly opened-up appointment slots is a significant challenge as well.

As a patient, or someone needing an appointment, one of the greatest frustrations is the wait for an opening, yet openings happen all the time when appointments are cancelled, rescheduled or no-showed.

Schedgehog.com is a unique and inexpensive web-based application, free for patients, that connects patients desiring same day or same week appointments with recent openings in physicians' offices by making the appointments visible as they become available on both a website and, importantly, on a mobile device like an iPhone, Droid or other smartphone.

When an appointment opens up, your receptionist enters it into Schedgehog's database. If a patient realizes they have an hour or two gap in which they'd love to knock out a doctors visit, they go to Schedgehog on their mobile, type in their doctors name and, voila! A list of all the appointments available same day at their doctor or within a certain radius of their location appears. One click and they are on the phone scheduling their appointment, keeping your schedule full.

I created the software behind Schedgehog as a way to recover lost revenue at my private optometric practice, Shady Grove Eye & Vision Care. I hired a team of web designers and after several iterations, came up with a web based mobile application. Subsequently, we sent letters and emails to our patient base to inform them that the practice offered this value-added service. One patient suggested that we open up the data base for all medical professionals and Schedgehog Professional was born.

For just $390 per year per physician, Schedgehog can plug up a $32,000 per physician drain on your practice and at the same time enable you to offer a value-added service of convenience to your patients who value your time and would love a chance to be able to see you without having to wait two weeks or more. For most doctors, just one or two open appointments filled by Schedgehog.com pays their entire years subscription.

What does this mean to larger practices with 10 physicians or more? You do the math. Sign up for Schedgehog with the code "Searchial" before the new year and receive 6 months free, including Schedgehog support. Email questions to schedgehog@gmail.com and connect at twitter @ Schedgehog.

Schedgehog.com also offers an affiliate program, where you can earn 25 percent of the recurring membership fees paid by those you refer.

CHAPTER 8:
SOCIAL BROADCASTING (LIFESTREAMING)

There are a number of mobile, live-streaming software applications developed and in development, and these applications are gradually enabling people to better upload and share videos from their cell phones to the Internet. These videos are shared on social networking sites like Twitter, Facebook, and YouTube. This movement is known as Socialcasting. Socialcasting is defined on Wikipedia as, "A movement in online video that combines traditional media content, social networking, and interactive community to create a unique experience for viewers on the web." Socialcasting developed out of several technology trends, including instant messaging, videoconferencing, social networking, video sharing, and blogging, and includes such sites as PodCasts, Ustream.tv, Justin.tv, and Skype.

Podcast

A podcast is a series of digital, audio, or video media files that are released episodically and downloaded onto a computer or mp3 player. A company that distributes the "file" of the podcast maintains it on their server, and the listener or viewer uses special software, known as a "podcatcher," that can access the file feed, check for updates, and download new files as they are generated. Files are stored on the user's computer or mobile device, providing convenient and easy use online or offline.

The name podcast implies a relation to the iPod product produced by Apple Inc, although there are many devices and forms of portable media players that podcasts can be used on, and any computer that plays media files can access a podcast. Some use the term netcast instead of podcast, to avoid suggestion that the "pod" is part of Apple Inc. Some suggest that PODcast is an acronym for Personal On Demand broadcast.

There are hundreds of businesses offering podcasts, and they are easily searchable on the Internet by Goggling podcast, plus a search keyword of interest. A podcast is an excellent way to build a following locally; be sure to advertise it in your business so your clients see it; put a link on your blog and website and regularly tweet a few days before it occurs. It is also a great way to build your reputation as a thought leader in your respective field. Podcasting is easy and intuitive. The links can be placed on different dofollow social sites to provide link juice, making it searchial for the content you post.

Ustream

Ustream is a platform for Lifecasting and live video streaming of events online. The website has more than two million registered users who generate 1,500,000-plus hours of live streamed content per month, with more than ten million unique hits per month. By using your desktop or laptop and internal or USB camera, you too can produce your own broadcast stream.

Ustream evolved out of the founders' desire to provide his deployed army friends in Iraq a way to communicate with their families back home. It provided a way for soldiers to contact family members all at the same time when their ability to make phone calls in the war zone was limited.

Users can broadcast from the website or a mobile phone. Users can watch streams on the website, iPhone, and Android applications. You can use Ustream as a promotional tool for your business or to enhance your reputation as a thought leader in your particular area of interest. Your stream should be promoted beforehand via your social media channels to maximize your audience. Again, create your video and proliferate the links socially that drive people back to your content and allow the search bots to find your efforts. Posting your Ustream video links within your blog or website and tweeting about them can help raise the searchial profile of those efforts.

Skype

Skype is software that allows users to make voice calls over the Internet. Calls to fellow Skype users are free, while calls to landline phones mobile phones are charged using a debit-based user account. On Skype, one can also take advantage of instant messaging, file transfer, and video conferencing functions.

The name originated from the original name of the project, "Sky peer-to-peer," abbreviated to "Skyper." As the domain name was already taken, they dropped the r and were able to get the domain name Skype.

You need to register to use Skype, are given a Skype name, and are listed in the Skype directory. You can communicate with other Skypers via voice chat and messaging. Skype offers a wi-fi phone that allows users to make free Internet calls to anyone with Skype anytime there is a wi-fi Internet connection.

Use Skype for conference calls or communicating when you might otherwise incur long distance or overseas charges. You can put a Skype widget on your website to enable people worldwide can contact your business for free. Suppose your patient is working in Japan and loses a contact lens—think how they would appreciate a toll free call to ensure their contact lens will be ready for them upon their return. I know they could e-mail or text, but there is nothing like the soothing voice of your receptionist to reassure them. By inserting a Skype icon on your website, you can get a searchial bump. Generally, Skype's best business advantage is to make business more convenient for your customers. Give your customers the opportunity to connect with you any way you can outside traditional means. We have several patients who serve in the military overseas and need to order contact lenses. Skype is the perfect tool for adding convenience to their experience engaging with our practice.

Video-Sharing Websites and Socialcasting

A video-hosting service is defined on Wikipedia as, "A service that allows individuals to upload video clips to an Internet website." The video host will then store the video on its server and show the individual different types of code to allow others to view this video. The website, mainly used as the video-hosting website, is usually called the video-sharing website.

Dr. Alan Glazier

YouTube is a video-sharing website. The popularity of blogs, forums, and social media has launched the utilization of video sharing services through the roof in the past few years. Video sharing websites are evolving; at the time of this writing, YouTube, the most popular video-sharing website, evolved into the third-most-popular search engine, behind only Google and Yahoo!. The increasing use of mobile technologies that incorporate cameras, both still and video, will be a huge factor in the evolution and growth of video sharing moving forward. The easy access of these tools has spawned a huge industry around the sharing of images on the Internet. Images from these devices are easily shared as they are typically low-bandwidth, low-resolution images that are more easily shared than DVD movies and other media that require too much bandwidth. There is an enormous demand for user-generated video content. For anyone who has used YouTube, you have to admit, the world is a much more interesting place since the advent of mobile camera technology and video sharing websites.

With the increase in usage of PDAs and smartphones, we bear witness to the early stages of the mobile web, and mobile video, just before the enormous impact they are sure to bring. Right now, mobile video has challenges due to the relatively slow connectivity and trouble handling too much bandwidth, but it will evolve parallel and out of existing technologies. In the not-too-distant-future, we will be able to effectively use our mobile devices to view and deliver higher-bandwidth, higher-resolution images like the ones we are used to watching on our desktops.

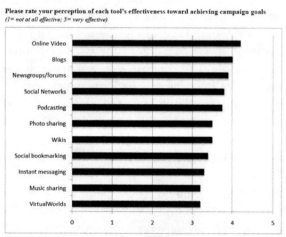

source: http://www.marketingvox.com/social-media-marketing-still-lacks-strong-metrics-037724/)
credit to Society for New Communications Research http://sncr.org/

Practical Marketing Applications for Video Sharing

Video is pervasive on the Internet. You can't do a Google search without pulling up thumbnail image or video search results. As recently as two years ago, this was not the case. We are just at the beginning of a huge wave of Internet video applications and usage. Incorporating video into your web efforts is not only important but is quickly becoming necessary, especially for searchial. Websites without video education tools, video descriptions, or other creative uses of the medium are viewed by the younger generations as old, out of touch—"so web 1.0."—Even more importantly, search bots reward you when they find subject-relevant video links on your website by giving you a bump in the search engine rankings. This is the most important reason to get video on your site. Without it, your competitors utilizing video on their website have a distinct advantage over you.

If you have yet to incorporate video, or have some videos either on or off your website but are unsure of what else you should be doing, here are some recommendations:

Adding video to your site can dramatically improve your ranking for some results, if done correctly. What search engines cannot do is determine the content within a video and use that information to increase rank. There are technologies in development that will use voice recognition and image recognition to make this possible, which is "so web 3.0," but at the time of this publication, search engines can't use video information as a ranking clue. This will be a highly important aspect of searchial marketing in the not-so-distant future. This includes the ability to provide tagging relevance, as the bookmark sites offer.

Tip: Go to Google's Webmaster Center, and you'll find information on how to tag and arrange your videos so that search engines better recognize them, and they can successfully be indexed. Be sure to give your video filenames that assist in relevance-based searches. Caption the videos correctly and submit a correctly formatted video sitemap that includes keyword tags relevant to your subject matter.

Ideally, one should try to put video within the body of every important page in your website, as the bots look favorably on video. There are some aspects of video that might cause people to leave your website early, which is undesirable; not only do you want people to visit your website, you want them to stay as long as possible. If you use too many graphics, it may take your website longer to load, reducing the chance visitors will take away anything from your efforts that will ultimately help your business. In using graphics, be sure to use descriptive alt attributes (see Chapter 6 regarding alt attributes). For example: <img src= 'new-contact-lens-picture.jpg' alt= "New Brand of Disposable Contact Lenses"/>. You can use these alt attributes, and even the image names, to increase the keyword density on the page being searched.

Your video images should have a "share," tell-a-friend, or other icon so visitors viewing your videos can easily proliferate the link, driving more visitors to your page and your video. Be sure to upload the video into Google, YouTube, or any other video-sharing website, where it is sure to be watched and has a better chance of becoming virally shared.

Making Your Own Videos

The process of making your own videos is simpler than you might think. Most computers already come with the software you need to make a simple video on your site. For Microsoft users, the product is called Windows Movie Maker. Mac users have iMovie, which is an incredible program in terms of a small learning curve and ease of use. This gives you a basic video to start with and play around with until you learn more advanced techniques.

You can use these programs, (or others out there), and upload live video that you have recorded, or, if you are more camera shy, you can use free clipart like ones provided by Microsoft or many other locations on the Internet. Video can be powerfully effective when added to your site. They give it a personal presence, which is important to any website. When you are known by only a URL or name, your picture, signature, or a video with your voice adds an element people can connect with and will increase your opt-in rate (the amount of people who choose to download or "click" on your content), conversion, and traffic.

Steps to Producing Video

1. **Invest in a digital video camera with Microsoft Movie Maker or Apple iMovie.** Making high quality, professional videos is so much easier than you think—Microsoft and Apple both have software that allows you to download video and stills from your hard-drive camera

directly to your computer, edit it, add effects and sound, and produce an incredibly professional video for free. I'm telling you now that even for the most technically challenged reader of this content, the software is *incredibly* intuitive and easy to use. I am including this link to a practice video I posted on YouTube recently so you can see the practice introduction video I made for my private practice after twenty minutes of filming and twenty minutes of editing using Microsoft Movie Maker for the second time in my life. I have had more than three hundred visits so far. Here is the link: http://www.youtube.com/watch?v=AKcnJnrmSiU

2. **Talk to corporate reps, find out the types of videos they produce that you may obtain for free, and post them to your website.** This is a cost-effective way of quickly getting relevant videos on your website, providing you a boost in search engine rankings.
3. **Hire a professional videographer to film your video idea.** Last year, I formed a charitable effort, which I promoted on Facebook and Twitter, where we provided free eye exams to people in need. For every one hundred people who join our Facebook group Shady Grove Eye Care—Just Joining Provides Free EyeExams to Those in Need, we provide a free eye exam. We promote the effort via social media, building goodwill with the local public, giving people in need care without need for insurance or exchange of money, and marketing for our practice at the same time. This is the professional video effort I produced: http://www.youtube.com/watch?v=ZC47TGfUfUE. I think it is quite compelling. Since we shot this, I now have three additional patients we are providing free eye care for, and we are shooting videos like this one, to post. As of today, one year after starting the effort, I have more than a thousand members in that Facebook group and will have given away a total of ten eye exams. Remember, in the social web: give back and they will come. Consider starting with common things, such as a how-to-host-a-New-Years-Eve-party video, a welcome-to-our-business video, or a meet-the-chefs video, to promote your catering business. If your blog is linked to your website, you can attach the video to the blog for free and benefit from the bump in searchengine relevance the video provides to both, although it is better to have the video incorporated within your practice website. If you don't have a blog or are in the process of getting one, you can hire your website developers to incorporate video directly onto your website. It is worthwhile to get one or two videos up right away, even if you have to pay for them.

Dr. Alan Glazier

Promote Videos You have Made over Your Social Media Networks and Blogs.

If you haven't or decide not to link the videos to your website, post them on your Facebook page, post a short link to them on LinkedIn, and tweet about your posts on Twitter. Get the word out and get people to view your videos the more viewers, the more relevant your video becomes on the Internet, and that relevancy trickles down to benefit your other web efforts the video is linked to.

CHAPTER 9:
THE MOBILE WEB AND LOCATION-BASED SOCIAL NETWORKS (AKA PLACESTREAMING)

The fastest growing areas of Internet usage involve mobile devices. Mobile Internet refers to accessing the Internet from a mobile device, such as a smartphone, PDA (personal digital assistant), or laptop. Your customers are doing it; in fact, they're often doing it while meeting with you (don't you love that?), and it's likely you're doing it too.

Mobile Applications

Mobile applications refer to programs that run on mobile devices (as opposed to programs that run on desktop or laptop computers). You are likely to find fewer videos or images on mobile devices than on your desktops or laptops because smartphones and other mobile devices have slower Internet connections. While advertising is pervasive on Internet websites, the bandwidth needed to show banner ads and other types of Internet advertising makes it difficult to display them on the mobile devices, so most mobile devices are refreshingly ad-free (for now).

Mobile applications can provide information to friends, businesses, and networks. While such applications create a highly targeted marketing opportunity for retailers, they also provide increased social connectivity. They enable people to be more aware of things located nearby that would otherwise likely remain hidden. It enables people to become aware of opportunities that may save them time and, ultimately, money.

Mobile devices, specifically smartphones, have navigation tools, location-aware applications, or location-aware technologies built in to help navigate our environment. Location-aware technologies you are likely familiar with include GPS, A-GPS, CDMA, and wifi. Mobile devices, by nature of their mobility, come in handy as navigation tools, and location-aware technologies are used to help identify where these electronic devices are located. They also allow users to exchange information through location-aware applications that relay via location-aware technologies.

These technologies are being used by individuals, small businesses, retailers, major corporations, and many others. Universities use them for everything from locating sites, students, and resources around the campus, in urban planning, and even in archaeology courses. For example, imagine an archaeology course on site where a student makes a discovery of an important shard of pottery. The student takes a picture of it with a smartphone, which tags the geographical coordinates of the shard and automatically logs it into a database. With the click of a button, the important archaeological discovery is cataloged by location, identified in a photograph, and tagged with any notes the student hopes to add regarding the finding, such as date, time, etc. Future urban planning projects considering digging near the site will find via that same database that the site has archaeological value and will be able to avoid disturbing it. Other archaeologists will be able to use the data to study the site and subject matter. This is just one of thousands of examples of how location-aware applications can be used to benefit society.

Social networking will fundamentally change as your device discerns who is around you. Many of the most popular applications on mobile devices are used to influence the consumer-retailer relationship. Consumers are using social-networking applications to find retail items they seek, locate deals, preview menus before they get to a restaurant, read reviews, and take advice from their networks. Retailers are using it to push ads that target their marketing toward people on mobile devices within striking distance; they are also buying and collecting data on their customers and the customers who are patronizing their competitors, as well as participating in networks to increase awareness of their brands.

Your business can participate like the many businesses already marketing in this new mobile sphere. It's likely more people in your geographic region are searching for services and products through their mobile device than are responding to radio ads or direct-mail advertising. Most of your younger clients are using social networking on their mobile phones, and the use of mobile phone applications is skyrocketing. Like social media and networking, marketing in this manner is much less expensive and more effective than expensive print, radio, or TV advertising, and yes, even Google Adwords. Your efforts in online searchial marketing will, to a large degree, mirror themselves in the mobile world. Right now there is nothing special you need to do other than ensure you are optimized on the mobile web via Google Maps (Google Places or Google Local—see next chapter), Facebook places, Yahoo! Local, and Bing Local. As of this writing, the best place to optimize with searchial is Google Local.

Placestreaming

Lifestreaming (see chapter 8) refers to someone publishing content by himself about himself—rather narcissistic, one might think. Placestreaming, on the other hand, adds a layer of information and communication that Lifestreaming lacks. Placestreaming apps enable us to become the publisher of our own *Consumer Reports, Zagat Guide*, and GPS all wrapped up in one. Think of it as location-based Lifestreaming with greater community engagement.

Some placestreaming technologies, such as FourSquare, have gamelike qualities to enhance the experience. As the game is played, the application acquires more data, becomes more useful, and makes new connections with people who have similar interests or experiences in order to enhance the whole social media experience for everyone participating. As a destination location, you don't have to play the game to reap business rewards, but you do have to add value to the game and you do have to sign up your business.

Here are a few notes regarding why you should consider applying these technologies to your business:

FourSquare combines benefits of a review site like "Yelp" with gamelike qualities (see section on Yelp, Chapter 16).

- Businesses set up a page and sign in every day.
- Users "check in" when they enter a business that participates.
- People use reviews and comments others have left for participating businesses to assess the value of the service or goods being offered at that location.

Business Application: Get your customers in on the game; let them know you participate using internal marketing and by posting the FourSquare label on your website. Businesses may want to include giveaways and discounts (which add real value to the game for your customers) for those checking in from your location, writing positive comments after they've checked in, or using them as a platform to advertise your charitable efforts to.

Brightkite lets you check in at various locations in the real world to see who is there, who has been there, and who is nearby.

- You can stream from a place or about a place.
- It integrates Google Maps and Twitter, so when you check in, you can notify people where you are on both services.
- It is a great tool for busy parents who want their kids to know where they are or vice versa.

Business Application: Check in or post at any Starbucks during happy hour, and not only do you get a half-priced Frappuccino using Brightkite, but you unlock cool features and rewards. Once you've picked your favorite Frappuccino, it is added as a badge on your Brightkite profile for all your friends to see. These features are only available if you are checked in at a Starbucks during happy hour. Now extend this to your brand on a local level, sign up your business, and get creative with a Brightkite promotion—Starbucks is doing it, so think of the value it could bring to your establishment. Brightkite is also one of the first applications to use the hot new technology augmented reality, which alters your surroundings when viewed through your phone's camera.

Glympse

- Lets you share your dynamic GPS map location with someone for a preset period of time.
- Send a Glympse to your children when you're leaving work, and they can follow your movements in real time—gives a whole new meaning to "Daddy's home!"

Business Application: Human resources benefits come to mind when I think of utilities for Glympse. You can track employee movements and possibly use it as a time-clock function in some instances. Want to track your business associate as she arrives from out of town, or ensure someone is running on time to meet you? Glympse lets you do that too.

Google Latitude/Google Maps for Mobile is one of the most used apps out there. It allows you to see your friends' locations on a map or in a list.

- With Google Maps for Mobile, you have access to GPS routes
- You can see your location on a map on your phone
- You can search for any business or category of business
- You can have turn-by-turn GPS navigation
- One of the most interesting abilities is to view your information in layers. For instance, you can pull up your geographic location on a map, get a Wikipedia description of it, and find the best transit route from place to place
- You can also program it to alert you when you are near someone you want to know about; great for autograph stalkers, or just stalkers in general.

Business Application: You want to make sure your business is listed on all of Google Map technologies. People will be looking for you; you'd better forget about that *Yellow Pages* listing and get on the map. By participating in Latitude, you also get search engine elevation benefit in Google Local, making it a searchial app.

Additionally, several Placestreaming sites incorporate Yelp-like review benefits with location-aware technologies:

- **Buzzd.com**—the first location-aware site aggregator
- **Loopt**—lets you locate your friends, discover what they're doing, and message them. You can layer it with information from Zagat, Citysearch, Bing, and other ratings sites, and you can update your Facebook friends and tweet it out.
- **iWant**—is essentially the combination of three location-based apps: restaurant reviews (Yelp, UrbanSpoon), movie show-times (Flixter, Fandango), and *Yellow Pages* (Yellow Pages, AroundMe). iWant uses GPS, so you never have to enter your address or zip code. iWant is currently available for iPhones only.
- **Yelp** has recently added gamelike qualities in order to compete in this arena, specifically with FourSquare, via an iPhone app. Twitter made a recent acquisition (Mixer Labs) that reveals a strategy to integrate

location-based service too. Facebook then followed suit with Facebook Places.

It's not difficult to visualize where this is going; with the addition of augmented reality to placestreaming apps, you will walk through a shopping mall, point your phone in the direction you are headed, and virtual conversation "bubbles" and banners will guide you to the products you want and the discounts and specials in each store. It stands to reason you likely will be advertising this way for your business as well.

Google Local (Google Places)

Google offers an incredibly useful free business listing called Google Places. The service enables you to advertise a coupon for free for unlimited impressions of your ad, and if you choose, enable viewing on mobile devices so your coupon shows in searches that occur through geolocation technologies, such as PDAs, or on placestreaming sites, such as Foursquare.com and Gowalla.

In order to participate, you need to log in to your Google account and find the home screen for Google Places. Sign up and fill out the requested information on your business. I advise you to incorporate a coupon of some sort and not just write an ad, as people within Google Places and Placestreaming in general are more apt to respond to incentives. To do this, go to Dashboard and find "coupons" to design a coupon for services offered at your location.

An important thing to note in designing your Google Places ad—Google gives you the opportunity to describe your products and services in great detail and for free. Don't limit yourself by leaving out attributes about your business. Tell potential new clients what sets you apart; don't be shy. Be sure to include links to your Facebook fan page and your Twitter account.

Google Places calls adding things like coupons, video, etc., tagging. Be sure to tag your Google Places ad with photos and video if you can, as these help raise your profile within Google Places, though not within search in general.

If you add information about your local business listing at Google Places, you might rank higher than your competitors. Google shows map results ahead of the traditional organic listing, arguably the most important real estate on the Google search page. The service provides free traffic reports and other benchmarking assistance. If you choose to use Adwords, you can link your Adwords account with your Places account, and your text ads will show an address, making your ad stand out from your competitors'. The topmost recognizable free search position on most search result pages in a Google search is Google Places, or Maps. The business listed in the A position has a tremendous advantage. By filling out your profile and adding video and pictures, you can increase your position relative to other businesses that don't fully take advantage of the Maps profile page.

At the time of this writing, Google Maps, Places, or Local, or whatever they are calling it tomorrow, is the most important place to optimize on the entire search page. It comes up first, just below the top horizontal Adwords listing. Optimizing Google Maps with searchial drives *a lot* of business. It is a great place to focus searchial efforts first. Post videos, photos, hours of operation, everything they ask for; the more the better. Once you gain a good position in Maps, try not to change data like the URL, phone number, or anything else if you don't have to. I have heard that that can temporarily hurt your ranking in Places and you may end up ranking lower than you did in your original position.

I have heard that participating as a paid subscriber on various Google-owned Internet properties, and tying them to your Google Places listing, adds priority to your listing, helping to elevate you higher in Map listing, although no one seems to be able to verify this. For detailed information about setting up your Places account, visit http://www.GooglePlacesHelp.blogspot.com

CHAPTER 10: SOCIAL REVIEW SITES, SOCIAL INFLUENCE, LOCAL SEARCH, AND CUSTOMER SERVICE TOOLS

Social Business Reviews and Online Reputation Control

The following was taken from the book *Groundswell*, by Charlene Li and Josh Bernoff: "Pilot Gabrielle Adelman and photographer Kenneth Adelman decided to photograph the entire California coastline (see their work at www.califoniacoastline.org). Singer Barbra Streisand insisted that photos of her house be removed, which was about as effective as trying to get rid of a hornet's nest by hitting it with a baseball bat. Of course, the resulting publicity caused people to copy the photo and post it to sites all over the net, easily found using Google Image Search of 'Barbra Streisand house.' Mike Masnick, a blogger for Techdirt, coined the term 'Streisand effect' for events where attempts to remove content from the Internet cause it to spread broadly instead. So not only is Barbra Streisand's house still visible online—now her name has become synonymous with futile attempts to remove content from the net."

Something is happening online that you need to be aware of. People are talking about you and your business using social networking and review sites like Yelp and Yahoo! Local. Sitting on the sidelines, not wrapping your head around it, and refusing to pay attention to what's being said about your business will hurt your marketing efforts … not to mention your *bottom line*.

If someone posts something negative about your business, you *can* defend yourself. You might never be able to get the comments retracted, but here are the dos and don'ts of how to handle the situation:

- DO contact the person and attempt to fix the problem, just like you would an unhappy customer in your store
- DO dilute bad reviews with positive reviews from customers you know would give you a favorable rating
- DO comment alongside or respond to the bad review yourself, or with the help of a professional marketing expert, so your comment or reply is believable, fair, and well received
- DON'T be confrontational with the person who wrote the negative review
- DON'T be too aggressive in pursuing the person to remove the review; if he or she doesn't want to talk about it or consider changing it, leave it at that
- DON'T think negative reviews are always bad for business; sometimes they add credibility by increasing the perception that your reviews are real and not made up by you, your family, or your friends

Remember, you are the only one who is going to be able to defend your reputation, enhance your reputation by interacting socially in this new environment, and maintain your Internet presence professionally; so take advantage of as many opportunities as you can.

The world of social media is an excellent place to do your own PR, and sometimes that involves promoting yourself, and other times that involves defending yourself. The most important thing to remember is to take the high road when defending yourself against bad posts, adopt an I'm-going–to-fix-this attitude, even if you have to eat a little crow. You don't want some "effect" named after you!

Before social media, one unhappy customer would maybe tell ten people about a bad experience; now, one unhappy customer can spread venom to *thousands* of people about your business easily with the stroke of a keyboard and click of a mouse.

It's a new world out there, but some of the same tenets of high-level customer service still apply—you can't take down a bad post and you can't retract a bad response, so be careful what you say, use kid gloves, and, if you ultimately can't make it right by them ... dilute the bad review as much as you can with positive reviews.

More and more search engines are able to return local results for keywords and keyword-related phrases. This is a particularly important technology for businesses. Not only are people looking to find businesses that are in close proximity to them, it is helpful to get ratings and reviews of the service providers in the same click, hence the popularity of the "local search" sites, such as Yahoo! Local, Google Maps (also known as Google Places), Yelp, and others. The increased prevalence of location-aware apps is also adding to the depth of this tool in a major way. Optimizing your site for local search and making sure you're listed is imperative, quick, and easy to do. So what are you waiting for?

Take the time to add as much information about your business when you sign up on these sites—most of the information can be added for free, and it is silly not to maximize your advantage, especially when your competitors are showing up above and below your listing and may have better or more information for potential customers.

Be sure to make sure that your address is inputted correctly. Humans may be able to interpret a poorly written address, but the bots that put you on the 'map', especially the location aware Google maps that people frequently refer to might create the wrong 'geocode', causing customers to miss you when they can't easily find you on the map.

Google Places uses a "location prominence score" to "locate" you on their maps relative to your competitors. This score is based on more than just your location. It is based on:

- A score associated with an authoritative document
- Total number of documents referring to a business associated with the document
- The number of documents with reviews of the business
- The number of information documents that mention the business (such as CitySearch and Zagat.com)
- Numeric scores of reviews
- Type of document containing the review (e.g., a restaurant blog, CitySearch, etc.)

Geographical location may be determined from different vantage points; centerpoint location is when the algorithm uses a "place" within the user's vicinity as the "center" and marks locations based on their proximal radius to that central location. Some geographical areas are determined in algorithms by zip code and use a second function when two businesses are dead center within a postal code. Other areas are defined by latitudinal and longitudinal coordinates, a more popular method since the advent of GPS and geolocation services. When this method is used, your ranking is based on two or more factors, one likely being relevance. There are several other algorithms used, such as search area, map boundaries, and z codes, which are a set of codes used in certain radio communications.

Here are a few tips for optimizing your position in local search: be sure to embed Google Maps on your landing page. Not only does it add to the interactivity on your website and enable customers to map your location, it counts toward your relevance ranking in Google Places. Always include driving directions; not everyone is comfortable using GPS or geolocation apps. Be sure to manage your listing at Google Places at the Google Places Business center—just Google "Google Places Business Center" and log in. Optimizing for Google Places involves all the traditional elevation techniques, but for local search, you might need a local blog and offer coupons. Be sure to fill in the two-hundred-character description field on Google Maps for your business. Include hours of operation and some text that distinguishes you from your local competitors. Google Places allows you to load ten images and five videos; take advantage of this and don't be concerned about upload times. It only adds to your site's local reputation.

Yahoo! Local

To optimize your Yahoo! Local listing, be sure to have the best keyword in your business name on the listing. If you are an office manager in San Francisco, your Yahoo! Local listing should include the phrase "office manager San Francisco." The enhanced Yahoo! listings get better billing in Yahoo! Local, so be sure to upgrade. Enhanced listings let you add your logo, tagline, pictures, a more detailed description, coupon links, and up to five categories for your business. Spend time writing your description and include as many pertinent keywords and keyword phrases, as this will optimize your listing when scoured by the bots. The best thing you can do, however, is get good ratings from your clients. The better your ratings, the better chance your listing will get "hit" for new clients. This won't, however, cause you to float higher in the local listing, but people will be drawn to those with good ratings at a much higher rate than those with poor ratings.

In order to be listed on Yahoo! Local, you have to show the name of your business and contact information on every page of your website. You also are required to provide contact information, and it can take up to a week for your listing to appear.

Bing Local

You can't enter information on Bing Local using Google Chrome, you have to use Internet Explorer or Firefox, and you have to have a Windows account to sign in. You can get a free windows account by registering for a live.com, msn.com or hotmail.com account. Listing on Bing Local is free, and you can update your listing at any time. There is a verification process to protect your listings against unauthorized changes. Fill in as many of the items on the long form as possible, and pick as many of the six categories they offer as you can. Bing's search refinement is supposed to be the best, so picking these categories should help your "float" (higher visibility) results.

Yelp

A review site ever increasing in popularity, Yelp sets the bar in terms of social reviews. Yelp allows you to rate and comment on businesses you patronize. Yelp uses an algorithm that analyzes user feedback to increase your position on the Yelp search page for a particular keyword or key phrase search. You can proliferate your yelp reviews to your Facebook page, which has serious implications for businesses, as it opens up a new avenue of exposure for reviews generated on Yelp. If you want someone to "yelp" for you (e.g., rate your business), it helps if they have rated businesses before on Yelp—the Yelp algorithm gives preference to Yelpers who have rated other businesses. This helps keep people ratings more trustworthy. In our office, we have a computer in the reception area with the Yelp homepage up all the time. Our receptionist asks patients if they use Yelp, and if they do, we ask them to rate us right then and there. This helps increase the number of people who actually go to the site and rate us. In the near future, look for Yelp to become more than just a ratings site—it is becoming more of a social network of ratings, recently adding Facebook and geolocation features.

Your presence as a business on Yelp is crucial. Be sure to have someone monitor reviews, encourage clients to post reviews, and engage in the social aspect of the site. Be sure to put up Yelp signs in your place of business letting your clients know they are welcome to rate you on Yelp, and include your Yelp link or badge in appropriate patient social communications such as emails or on your website. Blog about your Yelp efforts and tweet about them as well.

On their website, Yelp describe themselves as "an online *urban* city guide that helps people find cool places to eat, shop, drink, relax, and play …" Yelp has a lot of traffic and garners much attention in urban areas such as San Francisco, New York, and Washington DC. Communities of Yelpers in these areas can do amazing things for small businesses through word-of-mouth reviews and events, but the value of seeking and obtaining Yelp reviews for business falls off precipitously the further your business operates from a major metropolitan area.

If you are spending your time seeking online reviews to enhance your businesses image in an urban, suburban, or rural setting, I suggest you spend your time and energy building your customer reviews on Google Maps (soon to be Google Places) and Yahoo! Local before you spend time generating reviews on Yelp. No matter which setting you practice within—urban, suburban, or rural—most of your customers who search online start by using a major search provider like Google or Yahoo!; a comparatively small percentage start out using Yelp, even if you are in a major city.

Yelpers tend to be young, urban, tech-savvy people under the age of thirty-seven, and you are not likely to draw well via Yelp if your target market is greater than the age of thirty-eight. Those people are searching for you within the major search engines. As things stand, Yelp is positioned to be a major player, so don't ignore Yelp; there are those who find you in the major search engines and check your reviews on Yelp, so be sure to build your presence on all the review sites you can, but don't forget to focus getting reviews on sites most people visit, regardless of your business setting.

Social Influence

Bing and Google recently stated that sharing links through Twitter and Facebook directly affect search engine optimization (SEO). While this was assumed to happen, no one at any SEO company would state that the engines used information in this way.

What Google and Bing are saying specifically is that when a link to an article is circulated, or in Twitter-talk retweeted, the search engines take note. Even more specifically, Google and Bing give preference to people with greater authority i.e. people who tweet, Facebook and use other social media tools repeatedly to discuss specific areas of content. These social media "influencers" are viewed by search engines as more relevant, and, as such, their content is given preference in the search, especially for those topics they are prolific in publishing about.

Google specifically stated, and I quote *"It [social authority] is used as a signal in our organic and news rankings. We also use it to enhance our news universal by marking how many people shared an article."* Another surprising bit of information is that Google and Bing both admitted to computing calculating the authority of a user and tying that to the quality of the information they post. Both Bing and Google calculate whether a link should carry more weight depending on the person who tweets it as well. Google calls it "Author Authority" and Bing calls it "Social Authority."

In Danny Sullivan's interview with Google and Bing, "What Social Signals Do Google & Bing Really Count?", (http://searchengineland.com/what-social-signals-do-google-bing-really-count-55389) he goes on to list important facts about how authority is used to "grade" content:
- **Diversity of Sources** – posting links solely from your blog or website isn't nearly as valuable as posting links from across the internet.
- **Timing** – there is value in re-sharing old content; it's not just new content that gives you authority.
- **Surrounding Content** – the content you provide in the tweet or post gives additional clues as to the relevance for any particular subject matter you are providing information on.
- **Engagement Level** – how many people share your information

This is a paradigm shift towards what is called "organic marketing" – as the internet evolves search engines are finding ways to pull up more relevant information for your query. This is beneficial for everyone, except the spammers, and I'm sure none of you are shedding a tear over their misfortune!

Klout

The leader in calculating social influence at the time this book was written, a website called **Klout.com** began integrating it's "klout" ratings with data it compiled from Facebook in order to get a fuller picture of one's efforts within the social sphere. They plan on interfacing with LinkedIn, YouTube and a host of other popular social platforms to increase their influence as the leader in rating individuals social influence.

The first place "Klout" will be important is your search engine ranking. Tweets from the most authoritative Tweeters are likely to make it to the top of search results first. How many people share the blog you post will count towards your total social score. Links shared by people with higher social scores will count more than links from those with lower scores. Keyword and key phrases placed in content, while important, will no longer be the gold standard in optimizing content. The biggest factor will be your participation in the important social suites, so I'm going to go out on a limb and try and coin a new phrase to live by in this strange new system – **"Get Klout or you're out."**

Customer Service and Collaborative Tools

Uservoice

Uservoice is a widget you install on your website. It allows your customers/clients to easily and quickly provide feedback on anything related to your business. Customers vote on their favorite ideas; the items that receive the most approval or votes move to the top, so you can focus your attention on these items to make your clients happy. You can respond to your customers' feedback, and you can engage them and let them know you're listening and that you care. Each week, customers get e-mailed with responses to the feedback you left, so they stay engaged with your business throughout the year. Excellent tool!

Hy.ly

A simple, attractive application for your customers to share private account data and issue details using Twitter. With Hy.ly, you can empower your team to monitor service requests by customers and streamline issue resolution. You can set up a method of "listening" when your brand is mentioned on Twitter—a notification system lets you track what is being said about you on the Internet. Keep messages unread across browsers and team members so your team will never miss anything. A formal ticketing system enhances accountability and measurability of your staff. Hy.ly allows you to mark customer sentiment using a keyboard shortcut or a simple one-click process. Messages can be tagged to identify customer advocates, critics, or influencers. Your customers have the ability to add comments to their "tickets." There is no character limit, the exchange is private between the patient and your company, and the exchange is integrated into the dialog view.

Internal social systems like Hy.ly require participation across the entire spectrum of employees—managers as well as employees—at the corporation. Otherwise, they just don't work. It is important to develop a culture of electronic communication within the business and to nurse it by participating within it. Implementing these systems isn't something that can just be delegated down the chain; it has to be implemented in a top-down fashion where everyone continues to participate. Chris Brogan, an Internet marketing thought leader, says in his book *Social Media 101*, "Look at your organization's information needs. Don't start by pushing social media tools down people's throats, but instead look for the problems different parts of the organization might need to solve."

Check out the Hy.ly facebook "Welcome" page application on my facebook page http://www.facebook.com/OptometristRockville?sk=app_109486035783059 . This easy to use, inexpensive and intuitive tool can enable you to have an excellent and sophisticated welcome page for new visitors you hope to have "like" your page.

Other Tools

Yammer, JiveSoftware, and BaseCamp can help your organization communicate internally and collaboratively. They are basically enterprise social tools. Yammer is easier to use than Hy.ly but offers less functionality.

Social eMail: The Importance of Integrating Social Media in Your Email Marketing

In 2009, Boston College suspended the distribution of email addresses to incoming freshmen in favor of using social media. When I first heard this, it made me realize that social media was a potential threat to email communication and it didn't take too much to imagine the demise of email in favor of social media web wide. As it turns out - email isn't going anywhere - but instead we are starting to see a chimera arise, a **social**

email that enables better engagement with customers and the sharing of email content by customers to potential new customers.

2011 will be remembered as the year medical businesses integrated social media into their marketing strategy. The combination of email and social media is a powerful one, and there are at least two businesses making inroads into eyecare with tools that make your email communications more social – **Demandforce** and **Smile Reminder**, both of which entered the eyecare industry after success in other medical verticals. Demandforce is offering 1 free month after you sign up if you mention this book.

What does a "social email" look like? It is like a typical email with links to your social efforts; envision opening an email and finding icons that enable you to connect with the sender through their various social channels such as Facebook and Twitter. "Like" buttons will also be included, as will links that encourage the reader to write a review on a popular social review site like Yahoo Local, Google Places or Yelp. Social emails will contain incentives to share information about the sender with their friends inside and outside the social space.

What does this new technology bring to your business in terms of returns? You will be able to measure responses to your emails and the social links they contain as well as elicit feedback from your clients for business improvement and benchmarking purposes. This provides small businesses a **"Social ROI"** – a return on your investment in adding social technologies to your email, making it worthwhile to implement. Your consumers will appreciate the ease with which they can make suggestions that may make a difference in their experience at your business.

A study, *"Social Media's Influence on E-Mail Marketing According to U.S. E-Mail Marketers"* (http://www.emarketer.com/Report.aspx?code=emarketer_2000643) was conducted in August 2009:
• 81% of respondents agreed that social email *"extended the reach of email content to new markets."*
• 78% believed that social email *"increases brand reputation and awareness."*
• 53% believed social email *"increases the ROI of email programs in general."*
• 47% believed it *"accelerates the growth of e-mail lists"* and
• 31% believed it *"generates more qualified leads."*

No matter what one believes, the face of email is changing. Consider how your emails appear to your customers and patients compared to the colorful, interesting email with surveys and review opportunities they get from your competitors and ask yourself if it's time for your email marketing to evolve too?

CHAPTER 11:
ANALYTICS—BENCHMARKING YOUR SUCCESS AND MONITORING COMPETITORS' EFFORTS

Grading services

AlexaRank is a highly respected tool utilized by everyone from small Internet businesses to Fortune 500 companies. It is a listing of all the sites on the web, sorted by traffic. Alexa.com computes the reach and number of page views for all sites on the web on a daily basis. The rank of a site as calculated by Alexa reflects both the number of users who visit that site and the number of pages on the site viewed by those users. Alexa offers a toolbar you can install for free. The toolbar collects data on browsing behavior, which is transmitted to the website, where it is stored and analyzed. Alexa allows you to search its database, which helps you have your website or blog ranked and benchmarked for performance. Alexa crawls web pages that are submitted to it, not unlike the way major search engines crawl the web. It collects data from its crawls and uses it to enable users to compare their website's benchmarking to other websites. You can list your website with up to three competitors' websites, and AlexaRank will generate a graph that compares important website benchmarks to compare your web efforts to those of your competitors.

Hubspot.com

Hubspot provides tools to grade your Internet effort on many levels, from website and blog, through social-media efforts like Twitter and Facebook. On any of their many graders, you are prompted to enter the domain URL, and in a few seconds, you will receive a full analysis, including explanations of each result. The tool provides a grade between zero and a hundred based on the criteria it analyzes. The top of the report will show the grade with a basic explanation of what it means.

While most people find the tools helpful and relatively consistent, many people criticize the consistency of the results. I think for a free tool it's not bad at all. If you run your website or blog, be sure to include the "www" before the domain, as this is frequently a reason for finding less than-true results or inconsistencies in the results the Hubspot tools bring up. The free tool is a good guide to follow if you are starting, but I don't recommend buying their product. By implementing the advice in this course, you can greatly increase your search engine ranking and monitor it on AlexaRank for free, which tends to be more consistent in its output. As you make changes, you can watch your rank via AlexaRank increase linearly, yet Hubspot's tools can be all over the place. Hubspot offers many tools that look exciting at first. Who wouldn't want to know how they're doing on Twitter, Facebook, or on their blog? Below are several sites run by HubSpot.com which offer a free (subjective) assessment of how your efforts are working within certain social media sites.

1. Bloggrader.com
2. Twittergrader.com
3. Twitter.com/grader
4. Foursquaregrader.com
5. Facebookgrader.com

Other websites offer to grade the performance of your entire website include Googlerankings.com and Quantcast.com. Performance grades are based on utilization of newer internet technologies such as forms and tags, technological sophistication, technological complexity, implementation of social tools and compare your efforts to those of top-tier websites.

Competitor Spying

Part of an online strategy may include benchmarking and comparing your web efforts to those of your competitors. Many call this competitive analysis "spying," which negatively connotes what is a perfectly acceptable way of measuring your success compared to your competitors'. There are tools that allow you to see which keywords your competitors are bidding on in Adwords, etc. Tools exist online with which you can learn about competitors' trends, products, and strategies they are using to market online.

Spyfu.com

With spyfu.com you can enter your website and the websites of competitors to learn which keywords they use and how effective they are. You can find out what their budget is on Adwords campaigns, what the value of their ads are in terms of return, and the kind of traffic they receive organically versus via paid advertising. It also will recommend a list of the top organic keywords that are searched, and who, based on these keywords, your most significant Internet competitors are. It also can tell you which other domains your competitors own.

Compete.com

Compete.com offers comparison of your website statistics to those of up to two competitors. It can compare the following monthly metrics between you and your competitors:

- Unique Visitors
- Visits
- Page Views
- Compete Rank
- Average Stay
- Visits/Person
- Pages/Visit
- Attention

And can compare the following daily metrics:

- Daily Reach
- Daily Attention

The free version will tell you how many unique visitors you receive per month, the change in volume relative to the previous month, and will compare your current stats to the previous year. I find compete.com to be accurate and a particularly useful tool.

Statbrain

Statbrain can delve deep into the backbone of sites to mine statistics for you, such as how many visitors come to your site through back links. It is a relative benchmark; i.e., it does not precisely show how much of your traffic clicks through to somewhere else. You can go to a competitors' sites to see approximately how many people who click on their site click through to yours

AideRSS

Use AideRSS to find out which blog posts on competitors' blogs are the most compelling to people searching for information in your field. You can use this information to come up with new content for your blog effort. Use the data to determine which subjects people are interested in on your competitors' blogs, but don't plagiarize them; come up with your own unique angle on the topic they posted. Just don't do this to mine! LOL.

FeedCompare

FeedCompare works with Feedburner. Feedburner is a tool that you can use to track your RSS subscribers—with FeedCompare, you can compare the size of your feed to competitors' feeds.

Xinu Returns

Xinu provides data on how your efforts for elevation are going in major search engines. They also provide data on how your social-media campaigns are running in comparison to up to five of your competitors. Other technical details are available as well.

Google Trends for Websites

With Google Trends, you can input topics to see how frequently those topics are searched on Google. It also benchmarks how frequently those topics have appeared in Google News stories and from which regions they are searched for most.

Google Insights for Search

More specific regional data than Google Trends offers.

Microsoft's Keyword Forecast Tool

This tool can take a suggested keyword and forecast an approximate number of how many impressions you can expect from it and provides some demographic information based on the distribution of those keywords

Microsoft's Search Funnels

Query Chain is a phrase used to describe how keywords may be typed into a search query in sequence. Search funnels help visualize and analyze your search sequences, and you can use that data to input better search sequences and use the better sequences as anchor text and link text to help increase relevance.

WayBackMachine

This tool allows you to view your competitors' site over time and watch its evolution. If you think about it, the things they are doing that work are the things least likely to change over time, so you may want to take some clues for your web effort from them.

Web Page Speed Analyzer

You can compare how fast your web pages download to those of your competitors. Download speed is important—the longer it takes someone to download your content, the more likely visitors are to jump to a competitor's site. The faster your page loads, the more you typically convert visitors. Two sites that analyze web page speed are WebSlug and WebWait. WebWait is a better tool, as it takes into account script processing and image loading, which the other tools ignore.

Google Alerts

I use Google alerts every day to monitor my own efforts as well as those of my competitors. With Google Alerts, you can plug in any phrase, and when new content with that phrase appears in the Google search engine, you get an e-mail alert. You can set it to alert you daily, weekly, or monthly. Start by just inputting your name; that way, whenever something on the web or social media happens that includes you, you find out immediately and can use that information to read what is being said about you, or to thank those who are giving you publicity. You can input competitors' names, employees' names, or anything else you want to receive information on.

You can also set alerts for topics that you want to blog about, or to get notified when keywords that you use in your elevation efforts show up elsewhere online. You use that information to try to gain links from those sources as they are keyword relevant, and linking to them can boost your ranking in search engines. I have my alerts set on "eye," "vision," and "blind," and I get at least one idea a day for a blog post just from these alerts. I also use these alerts to find recent blog posts relevant to my industry, to which I reply, inserting a link to relevant content within my blog or website, earning a link back to my website, which helps my PageRank.

Analytics are the most important tool you can use to build a successful searchial marketing campaign. I heard one lecturer on the subject say a social media campaign without monitoring analytics is like running a business without accounting—you can't do it and expect to be successful. Some of the free analytics tools I use are Google Analytics, Websitegrader.com, Compete.com, and analytics tools embedded in my Wordpress blog content management system. These tools allow me to view a broad swath of my efforts, understand what is working, what might not be working, and get a handle on things. They allow me to make adjustments in my strategy to draw more eyes while hopefully helping me to save time and increase efficiency. Analytics can help you understand how effective your paid advertising campaigns are. It's also the place to look to determine whether your blog or website is an effective marketing machine or a web 1.0 dinosaur that's rarely visited, except when someone's fingers slip on the keyboard and end up there by accident.

Google Analytics

The Google Analytics homepage provides several graphs within the dashboard, which provides an overview of the most salient statistics, including site usage, traffic sources, visitor overview, and page views for particular content pages on your site. The statistic side of this can get pretty heavy, so we are going to concentrate first on teaching how to use the information under "site usage," which are the easiest statistics to understand and the ones you can make the most use out of when optimizing your site for search.

On the upper right side of the graph, there is a tab titled "AllVisits." Clicking this tab allows you to filter your report into:

- All Visitors
- New Visitors

- Returning Visitors
- Paid Search Traffic
- Non-paid Search Traffic
- Several other options

All Visitors

This tab provides metrics across all visitors to your website, both new and old. The graph will usually show steep peaks and deep valleys; the valleys usually occur as you head into the weekend, where most web traffic slows down, and the peaks rise from Mondays through Wednesdays, usually with downturn starting sometime mid to end of the week.

When you are in the All Visits tab within the site usage area, "Visits" provides a graph that breaks down the statistics of visits for all visitors, such as:

- How many total visits for the time period requested
- How many visits pages were viewed
- Pages/Visit, which indicates the average number of pages a visitor clicked on
- Bounce Rate—the percentage of visitors who "bounce" away to another site rather than digging deeper in yours
- Average time on the site
- Percent who are new visitors—i.e., hadn't visited before

The following information breaks down the different ways you can adjust and apply sources for which you search within Google Analytics. You can search sources within the following metrics by looking in the "advanced segments" tab on the upper right hand corner of the gray bar on top of the dashboard:

- All Visitors (discussed above)
- New Visitors
- Returning Visitors
- Paid Search Traffic
- Non-paid search traffic
- Search traffic
- Direct traffic
- Referral traffic
- Visits with Conversions

- Mobile traffic
- Non-bounce visits

Traffic Sources Overview

"Traffic sources overview" breaks down where your traffic is coming from (e.g., people finding you through search engines, links from other sites [referring sites], or directly, as in typing your URL right into their browser or from a bookmark). You can dig deeper by clicking the "view report" icon on the lower left of the traffic sources overview to find out more specific information on sources, such as which search engines referred the most traffic; and which keywords were most frequently used to search for you. Dig even *deeper* by clicking the "view full report" icons under the "sources" or keywords section for further information.

Your top traffic source will usually be from the Google search engine, which will normally be responsible for more than 55 percent of the visits to your website. If this number is lower, you should check to make sure you are indexed correctly within Google, or that you are indexed at all. You can also use this strategy to see which search engines may not have you indexed effectively—or at all, for that matter. Your biggest site of referral should be somewhere on that list too—for me it is my blog, as I put a link from each blog post to my practice's web page. This helps me see how effective my blog is at driving traffic to my website and enables me to make strategic decisions to try and improve those cross links. Knowing which keywords or key phrases people use most frequently to arrive at your website is helpful, and I suggest you use these words and phrases to create additional content, with the keywords and phrases as anchor text and link within them, as they clearly have been successful at driving people directly to your site.

Map Overlay

Map overlay is a tool that enables you to see which areas on the globe your site is most frequently accessed from. The darker the country—color depends on which browser you use. On mine (Firefox, as of this writing), the darkest green colors represent the countries with the most people visiting youreyesite.com. Hit "view report" in the lower left hand corner again, and you get more detailed information about the countries people are visiting from. This can be broken down into greater detail by city, subcontinent region, and continent. If you notice the gray tab titled "site usage," beside it is another tab, "goal set," where you can configure goals to locate visitors from desired countries.

Visitors Overview

"Visitors" is a different statistic than "visits." One visitor can come back to your site numerous times and is only counted as one person toward the visits statistics. Your visitors number should ideally be smaller than your visits numbers. The visitors overview is located at the bottom left of the homepage and can provide data within the following classifications:

- Visits
- Absolute Unique Visitors
- Pageviews
- Average Pageviews
- Time on Site
- Bounce Rate
- New Visits

Further, one can view visitor segmentation (on the right side of the screen) and break down visitors by:

- Languages
- Network locations
- Browsers
- Operating systems
- Screen colors
- Screen resolutions
- Java support
- Flash

An interesting new feature is the Map Overlay, which enables geolocation visualization. Google's "DoubleClick Ad Planner" is a tool within that box that can use the demographics and behaviors of your website's visitors help write and place more effective Adwords and Adsense advertising, if you are using paid advertising.

Content Overview

Content overview shows you how many pages were viewed on your site (PageViews), how many of those views were unique visitors (never been to your site before), and again, what the bounce rate for this group was. A helpful statistic titled "top content" appears below. It allows you to see which pages on your website received the most views. This data is important for your linking strategy (see section on external linking in Chapter 8). You can also use this function to determine whether all your web pages are indexed in Google by clicking on the "full report" icon at the bottom. If you click the full report button and look to the right of the gray bar at the bottom, it will tell you how many pages people have clicked on. My website has 143 pages, and the full report shows that people have clicked on all 143 different pages, indicating all my pages are indeed indexed. On the left side, you can break down the statistics by:

1. **Overview**
2. **Top content**—which pages were viewed the most
3. **Content by title**—which titles are viewed the most
4. **Content drilldown**—gets you deep inside the site's directory to see how keywords and phrases are structured and acts as a tool when optimizing pages for search
5. **Top landing pages**—pages people first landed on when visiting your site
6. **Top exit pages**—pages people left your site from; they either got the information they needed or something they saw scared them away; gives you clues as to which pages may need to be improved or rebuilt to keep people there longer. Improving these pages should help improve the bounce rate as well.
7. **Site overlay**—very interesting tool to check out; site overlay pulls up the homepage of your site and shows you the percentage of visitors who clicked on each link on your homepage.

Google Site Search Reporting

Goals

Google describes goal conversions as "the primary metric for measuring how well your site fulfills business objectives. A goal is a website page a visitor reaches once they have made a purchase or completed another desired action, such as a registration or download."

Setting goals enables you to see the conversion rates of visitors and the monetary value of the traffic your website receives. You can set up goals in the goals overview section.

Custom Reports

Google enables you to take any of these functions and create your own custom report, a helpful tool if you have specific methods of benchmarking for your business or want to limit the number of pages you have to click on to get the data you need.

Websitegrader.com

Hubspot's Websitegrader is an incredible tool that is free and simple to use. They offer a paid service, but most of the benefit of what is offered can be gained by reading this book or other books on the subject and guiding your own web developer. On the Websitegrader.com home page, enter your URL with the www attached. You can also enter competitors' website URLs to do a comparison. You will be directed to a page that gives your website a grade. This grade is based on factors, which have been discussed in this book, that can help you have a more relevant web presence and explain which ones you have implemented.

Scrolling down, you can see where you grade high, and which factors you are missing that may help improve your score.

1. Blog analysis
    We have previously discussed the importance of having a blog that is attached to your website. A blog hosted on a free blog domain might have some points deducted, as some feel it is a disadvantage over owning a blog domain and having a private host.
2. Recent Blog Articles
    You earn points if you have a regularly updated blog, as opposed to a stagnant blog
3. Google Indexed Pages
    This number shows how many of the pages in your website Google has found. If the number here is less than the total number of your pages, you should identify them and get them indexed on Google (see Indexing, Chapter 6)
4. Readability level
    Measures the approximate level of education necessary to read and understand the web page content. Content should be made simple so that the majority of your target audience can understand it.

5. Optimization
    a. Metadata—checks to see if you have the right tags set up
    b. Heading Summary— images helping or hurting your being found
    c. Interior Page Analysis—checks internal page optimization randomly
    d. Domain Info—the older your domain, the better search engines like it
    e. MOZRank—the ten-point measure of global link authority
    f. Last Google Crawl Date—tells you when the bots last visited
    g. Inbound Links—IMPORTANT—this is who links to you and directly influences your float to the top of search engines
6. Promotions
    a. Bookmarking—how your bookmarking effort is helping you
    b. Link Tweet Summary—how many times links on your site have been re-tweeted
    c. Twitter Grade—how popular you are on Twitter—not always accurate on this site
    d. Google Buzz Count—Buzz is a social-networking and messaging tool from Google that gets you a few points if you use it
7. Convert
    a. RSS Feed—if you have one, great
    b. Conversion Forms—any forms on your website
8. Analyze
    a. Traffic Rank—based on Alexa, a popular ranking service, tells you how you are doing compared to the rest of the Internet.
    b. Score Summary

Hubspot (the company that published the websitegrader.com tool) offers other paid tools that can help with your efforts, but you may be able to do it alone with only the advice in this book and others. There are many products that compete in this space, but this book was written as a step-by-step guide to help you get started and learn how to keep moving through this fast-paced and confusing world of social and search elevation, so try on your own at first, and if you need help, seek advice from me or other colleagues who are thought leaders, and then you can look to paid tools for more help.

Radian6

Radian6 allows you to measure, analyze, and report on social-media efforts. They also provide tools that help businesses engage with their communities online. They provide sources for scanning blogs, video, forums, etc., for content you are searching for, and they aggregate what they find for easy reference for you. They can segment, filter, or parse your social-media data to view and measure it through dozens of different "lenses." They offer enterprise-management options. I highly recommend playing around with this tool—it adds incredible value to your analytics efforts.

Crazy Egg

CrazyEgg can provide blog statistics, such as who's clicking on your blog or website, where they are clicking, which geographical region they come from, which browsers they use, who the referring sites are, etc.

CHAPTER 12:
SEARCHIAL MARKETING AND EMERGING NEW MEDIA TOOLS

The future of searchial marketing strategy is, to some degree, uncertain. What we do see are new media platforms emerging that enable search engines another data point in their quest to improve their algorithm for providing the best search. To rehash what was said earlier, search engine companies are in business to make a profit, and it is a competitive business. Google, Bing, Yahoo!, and many smaller providers make a profit by drawing the most visitors to their search tools. The better search results they provide, the more eyes they draw; the more eyes they draw, the greater opportunity to earn revenue through advertising, including sponsored, banner, and pay-per-click advertising. The more junk a search engine pulls up when someone searches the term "doctor," the less likely that person will return to that particular search engine to find information in the future, and the less that particular company will be able to compete in the search engine space. So which new media tools are emerging that can help search-engine companies improve their algorithms?

Social Review

Google Local, Yahoo! Local, Yelp, and a host of other social review sites are playing an increasingly important role in searchial elevation. Search-engine and social-review companies share the same desire in terms of improving the quality of information available to their users. The more trustworthy a social review site is, the more it is utilized by the public. The better the algorithm is at determining which reviews are real and which are fake, thereby eliminating them, helps a social-review site improve, gain followers, and monetize itself. The search-engine companies, interested in providing the best information, are starting to use data from the most trusted social-review sites, as it helps them pull up the best information first for keyword and key phrase searches. While Google and Yelp compete as review sites, and I have even heard it suggested that Google suppresses Yelp reviews from its algorithm, it is likely time will change that. Or the two social-review giants will come to some sort of agreement where data from both review sites will be used as part of the algorithm. This will make it important for you to have ratings from your customers on as many of these sites as possible to improve your rank in search engines. The more high quality, positive, and detailed reviews your business has, the higher you will rank in searches of keywords and key phrases of the products and services you offer.

Intelligent Information Discovery—Question and Answer Websites

A hot trend and getting hotter in searchial is intelligent information discovery, which manifests itself on Q&A websites. These websites are not necessarily new but are being revisited because they are a powerful search tool made more powerful by social networks like Facebook and Twitter. Advanced features give them an edge over the traditional Internet "forum" way of having questions answered. Q&A has been described as a technology that offers a way to surface the best possible answers in the shortest amount of time, with the ability to verify the expert credentials of the person who wrote the answers.

While a tremendous amount of information exists on the web, you may have noticed that sometimes, when you have a specific question, you can't find the specific answer all the time. If there was an expert you could drum up socially, you would stand a better chance of getting better information. Search engines thrive as businesses by bringing up the best information the fastest, hence the recent increased interest in Q&A technologies by venture capitalists. The most important thing for a Q&A website to offer is to have each question page become the best possible resource for someone who wants to know about the question, a kind of mini-blog on specific topics. Eventually, when looking for information on the Internet, you'll see a link to one of these Q&A sites and will breathe a sigh of relief, because you now have a place you can go to find everything you need on the subject. Q&A topics will have an enhanced reputation over Wiki posts because the authors of the answers will be listed, and you will easily and efficiently be able to find out what their training and expertise is.

Some popular Q&A sites:

- Answers.com
- Google's Aardvark
- Facebook's Questions
- Apple's Siri
- Quora
- Swingly

Answers.com is now ranked in some studies as the seventeenth most-visited site in the US. The vast majority of Answers.com's traffic is to user generated Q&A pages. Yahoo! Answers gets even more traffic. Much of your potential market is already getting their answers from these sites.

Providing quality answers and links to relevant pages can help your business's social media efforts and search engine optimization by:

- Directing customers (and potential customers) to information about your product
- Connecting with people in your market, which builds your reputation, and generates leads
- Provide links back to your site. Some of these links are follow links, and thus also provide SEO value

As search engine algorithms evolve, Q&A participation is likely to increase in importance as a factor that determines your rank in search engines.

"Likes" and "Check-Ins" on Social Media Platforms

I found this quote in an article the day before I submitted this book to the publisher as a "final draft." "Beginning today," said Yusuf Mehdi, senior vice president of online audience business at Microsoft, "Bing.com will start displaying pictures and names of a searcher's Facebook friends who have either 'liked' a business or checked in at one via Facebook Places. In short, it's bringing a similar Facebook imprint in terms of look and feel to Bing that has for months been seen on other instant personalization partner sites—Yelp, Microsoft's Docs.com, Pandora, and Rotten Tomatoes." (http://www.clickz.com/clickz/news/1742246/bing-deal-extend-reach-businesses-facebook)

Businesses fortunate enough to have people who "like" them and/or "check " at their location will benefit with a search engine "boost." The reason these tools are growing in importance is they are an indicator of how "social"—thus involved—a particular business is online. As we have established, social interactions within new media makes businesses more relevant in the eyes of the search bots.

Mehdi goes on to add, "The CEO is awarding Facebook.com advertisers who have been spending money with the primary purpose of increasing their 'likers' base. To be clear, the more 'likers' accrued by companies, the more Bing searches for their brand will appear with Facebook-driven social context." (http://www.clickz.com/clickz/news/1742246/bing-deal-extend-reach-businesses-facebook)

Facebook is taking a broad step toward increasing the relevance of participating in its "places" platform by striking up a deal with Bing, where, for example, people searching for restaurants using the Bing search engine will be able to ping Facebook users who have posted Places reviews, to ask for their opinions directly. There will be an interface that enables Facebook users to turn on or off the option to respond to these pings, so don't worry about the privacy issue.

This, again, adds depth (in this case) to the Bing search algorithm; Bing is poised to milk the huge database of linking and rating going on within Facebook to their significant advantage.

AFTERWORD

An eagle was resting on a tree branch, doing nothing. A rabbit saw the eagle and asked him, "Can I also sit like you and do nothing?" The eagle answered, "'Sure , why not?" So the rabbit sat on the ground below the eagle and rested. All of a sudden, a fox appeared, jumped on the rabbit, and ate it.

Moral of the story: To be sitting and doing nothing, you must be sitting very high up.—*Anonymous*

There are two types of businesses: growing businesses and dying businesses. In our modern era, businesses that hope to grow must adapt to paradigm shifts. If you use traditional marketing and avoid using Internet technologies to market your practice, your competitors are gaining ground on you or may even be far ahead of you. They are building networks, enhancing their professional reputation, and digging their talons deep into the fabric of the Internet. They are spending less money, so they have extra discretionary income to develop their web efforts, distribute elsewhere, or take home. To compete, you need to be on the ground, even in the trenches, learning how to manipulate networks to your advantage. Very few of us sit up high like the eagle.

In 2006, I had a dying business—I had fewer new patients appointing every year, and my traditional marketing expenses were high and not yielding what I knew they should. By the end of 2007, I had cancelled all my traditional advertising, cancelled my Adwords campaigns, and started to learn and implement social technologies with help from my friend and patient, Shashi B, a Web 3.0 innovator and social media thought leader, the Internet "swami" at Network Solutions. This year, I benchmarked my efforts, creating a before, during, and after implementation of my social media strategy graph showing change in volume of new patients from the period 2005–2010. The graph was compelling. During the period between Q4 2005 and mid 2007, I showed a steady decline in new patient volume quarter by quarter, which deviated to a sharp up-tick in Q4 of 2007, when the changes I implemented started to take hold, significant growth through Q1 of 2008, and a steady increase ever since. My traditional marketing expenses during most this time? I sponsored two little league teams for $400 each and periodically hired a developer.

If you have read this book, you have already made a commitment to learn about these new technologies and should feel good about that—you are still ahead of most businesses. The game is on, but it is still early, so read about, learn from, and listen to the experts. Use new Internet technologies to communicate with, not speak to, your client base. Slowly delegate when comfortable and watch your new business increase. That is the way to a seat on a comfortable branch at the top of the tree.

ABOUT THE AUTHOR

Dr. Alan Glazier is an optometrist and founder/CEO of a large, private optometric practice in the Rockville, Maryland, suburbs of Washington DC. Dr. Glazier is the inventor of a novel intraocular medical implant, LiquiLens, capable of restoring near vision to people blinded by macular degeneration. The device is poised for its first human study in Europe later this year. Dr. Glazier has been issued patents and has nine patents pending on technologies in ophthalmology and computer science. Dr. Glazier is a consultant for two of the leading contact lens manufacturers, thereadedge.com, and Demandforce. Dr. Glazier is a frequent lecturer on social and new media, and he blogs professionally at SightNation.com and for Jobson's "Click" e-newsletter publication. He has been interviewed by *Entrepreneur* magazine, on Blog Talk radio, and by Network Solutions regarding his cutting edge use of social media. In 2010, Dr. Glazier was awarded "Best Use of Twitter" by the Northern Virginia Technology Council, the nation's largest technology council. He is also on the editorial board of *Optometric Management* Publication.

In 2006, Dr. Glazier made the decision to eliminate traditional forms of advertising and started the transition of his practice toward EMR and online marketing. His practice is one of very few that use no traditional marketing, except for the occasional recall postcard. Dr. Glazier conceptualized and brought to market Schedgehog.com, a free mobile app for patients, which connects patients with cancelled or rescheduled doctors' appointments, making the patients happy and recouping lost revenue for clinicians.

Connect with Dr. Glazier on Twitter @EyeInfo, or e-mail him at aglazier@youreyesite.com. You can also find him on LinkedIn and connect with him there. Read what's happening in heathcare social media by following the #hcsm and #hcm hashtags on Twitter.

Dr. Glazier's Social Landscape

Twitter: @EyeInfo

Practice Facebook Page: http://www.facebook.com/pages/Rockville-MD/SHADY-GROVE-EYE-AND-VISION-CARE/212943468734

SEO Facebook Page: http://www.facebook.com/pages/OnToptics/148592848508786

Schedgehog facebook page: http://www.facebook.com/pages/Schedgehogcom/131514453560289

Facebook Places: http://www.facebook.com/pages/Shady-Grove-Eye-and-Vision-Care/158306020870112

Facebook Charitable Effort: Shady Grove Eye Care—Just Joining Provides Free Eye Exams to Those in Need—http://www.facebook.com/group.php?gid=63386876962&ref=ts

Practice website: http://www.youreyesite.com

EyeInfo Blog: http://www.youreyesite.net

Sightnation.com Blog: www.sightnation.com/blogs/alanglazier

LinkedIn profile: http://www.linkedin.com/pub/dir/Alan/Glazier

Skype handle: drglazier

YouTube EyeInfoChannel: http://www.youtube.com/user/EyeInfoChannel?feature=mhum

Foursquare: http://foursquare.com/user/666244

Yelp: Practice Page—http://www.yelp.com/biz/shady-grove-eye-and-vision-care-rockville

Yelp: Personal Page—http://aglazier.yelp.com

SEARCHIAL FOR MEDICAL ORGANIZATIONS

Physicians and the healthcare industry, in general, have been slow to implement social media strategies. Perhaps it comes with the territory; for years, medicine has had its hands tied by HIPAA privacy rules and has had to walk on eggshells when interacting via any communication medium with patients. Perhaps it's the stigma the word "social" evokes when tied to healthcare, as in *"social*ized healthcare" and *"social*ized medicine." Perhaps the physician community has built up walls, keeping communication to a minimum in the interest of time management because of the volume of managed care patients they need to see to make ends meet. Maybe there is nothing inherently social about organized medicine after all; hospitals are perceived by the general public as cold and uncaring, and many MDs have a reputation for poor communication skills and being inaccessible much of the time.

Maybe it's none of the above, but it *is* time for healthcare workers to open themselves to the idea that social media is here and can be used to change these perceptions. Changing the professions will undoubtedly lead to greater change in the medical field as a whole. Not only can social media be used to market a business, it is a powerful reputation enhancer; it is a way to change preconceived notions by patients. By implementing a practice or hospital social-media strategy, patients may become better educated, thus, better historians. They may become, via doctors' preventive medicine recommendations, better educated, more loyal, and more faithful, thus more willing referrers of friends and family. One strength of social media is its effectiveness as a PR tool. Social media can break down walls and change perceptions, elevating healthcare and the perception of its providers.

It's time for healthcare businesses to modernize, using the new medium to share ideas—to break down the old-guard medical and managed-care model and build a new, more social, medical-community complex, one that benefits everyone, not just those out to make a profit. I believe that through this medium, medicine will be forced to change—for the better. I believe this change is happening now.

In most other industries, including service industries, social media already plays an important role in the day-to-day operations of companies. Implement a social-media strategy within your healthcare business on the micro-level. Stick with it for six months and see which changes you notice. I promise, your business will never be the same; it will be better and will start to take a totally new direction. It will gain in technological sophistication, and its reputation in the community will be enhanced.

When Mrs. Jones has stable A1C levels (A1C is a test of utmost importance to diabetics as a long-term measure of control over blood glucose) for the first time in years because you set an *automatic* tweet that reminds her daily to stick to your dietary recommendations and exercise, you will witness the power of this new medium. When your patient thanks you for using Schedgehog.com, a mobile app that connects those patients who want to get in to see you sooner with cancelled and rescheduled appointment slots (recouping *lots* of lost revenue for you in the process), you will witness the power of this new medium. When you find your practice enhanced through the sharing of medical information with colleagues on LinkedIn, you will witness the power of this new medium.

A revolution is defined in Wikipedia as "a fundamental *change* in power or organizational structures that take place in a relatively short period of time." Social media and new media is a revolution, and you are swimming against the tide if you are ignoring it and continuing to use traditional patient interaction and marketing/PR methods without taking the time to learn what others are doing with social media. Maybe something else will come along that will revolutionize medicine, but it's a mistake to ignore the revolution in medicine that is happening right now. It's not socialized *medicine,* it's socialized *marketing* that's changing the playing field—it's time for the healthcare industry to join the revolution.

APPENDIX A:
THE FDA AND MEDICAL INFORMATION AVAILABLE ON THE INTERNET

The US FDA once considered how it should regulate medical information on the Internet. That was 1996, two years before Google even existed. In November 2010, the FDA held its second hearing, asking the public for comments on the topic, with the intent of using the information to consider the possibility of regulating. It should be no surprise that the Internet world waits with bated breath for this meeting, and rumors abound.

The FDA will seek data about how users seek health information on the web and are interested in determining what influences types of imagery on the web have on consumer healthcare purchasing, issues related to side-effect reporting, relationships between corporate entities and healthcare workers in terms of potential regulation of grants, and whether or not ads distract people from understanding information related to risk of healthcare products. They are looking to regulate the message that comes across via social media as it relates to pharmaceutical and medical device products.

This meeting will no doubt not result in immediate changes in the way healthcare is marketed on the Internet; they'll absorb information from the meeting and take a few months to study the data and eventually get around to issuing some targeted guidance, likely two to four years out.

APPENDIX B:
SOCIAL MARKETING FOR MEDICAL PRACTITIONERS AND PROFESSIONALS

The majority of this book describes methods and tools anyone in any discipline can use to participate in social media for business, build loyal networks of connections, and improve their position in search engine rankings. Most of these methods and tools can be applied to business across the board. This chapter, and the chapter that pertains specifically to clinics and hospitals is designed to demonstrate trends and non-specific methods healthcare concerns can implement practically. The down-and-dirty technical how-to techniques follow the first two chapters.

A Manhattan research survey published in early 2009 reported that 60 percent of US physicians are either actively using social media networks or are interested in doing so (AmedNews.com, August 9, 2010 http://www.ama-assn.org/amednews/2010/08/09/bil20809.htm) So what is keeping the other 40 percent away? As of this writing, three are no guidelines for the healthcare profession on the use of social media, so naturally, clinicians are confused as to what constitutes acceptable standard of care and professional responsibility within social media conversations. It is obvious to most doctors versed in privacy regulations or concerned about getting sued that one should not make recommendations for medical treatment on social media sites, such as Facebook and Twitter. But what if the doctor who uses these tools regularly misses a message from someone about an urgent problem, doesn't respond, and the patient doesn't receive timely care? Another thing to consider is boundary issues—if you discuss a medical condition online, either in social media or other forums or discussion groups, and you are not licensed in the state in which someone uses your information, are there any potential license issues? No one knows the answers to these questions as of this writing. A justified fear is that what you type to patients on social media sites can be referenced and is a permanent record of advice you might have been better off not giving.

Despite all this, clinicians and allied health professionals continue to use social media in ever increasing numbers. The likelihood is social media is here to stay, so while some opt to be cautious, others are taming the beast and benefiting tremendously. If you are of the cautious breed, I recommend setting up personal profiles within social media, limiting your connections to friends and family; this way, you will be able to learn the tool and quickly adapt when regulations in the medical field take hold. When your patients ask if they can connect with you via social media, you can politely explain your reasons for not connecting with people outside your immediate circle without sounding like a dinosaur—by the way, a dinosaur to a twenty-eight-year-old patient is someone who doesn't use social media.

It is important not to get too close to patients with social media tools, or to give the impression that you are willing to discuss specific medical issues online. Providing an easy way for your patients to communicate with you or going out of your way to communicate the way your patients like is viewed favorably by your patients and reflects well on you, by making you appear up-to-date and relevant, but patients don't expect it. It benefits your business and your reputation to start your business efforts in social media by distributing content online within your area of expertise. Find your inner author and create a reference within your area of expertise, solidifying your reputation as a thought leader. This will not only draw people to your online efforts, but the relevance of the content on your particular subject matter will be recognized by the search engines, enabling you to have a better position in the organic listings when someone searches for doctors, PAs, and nurse practitioners in your area. This is a particular theme within this text, and the most important aspect of social media in terms of driving new business; one particularly applicable to the small medical business, clinician, or allied health professional.

Parents are looking for authoritative information on their children's health conditions, and, I believe, are drawn to clinicians who participate in social media, as it is "social," adding an element of personalization and trust to the conversation. Patients with terminal diseases are also looking to make a special connection with a physician and are likely to search within social media. At the very least, people try to find clinicians who are producing the best content and gravitate toward them and away from physicians who appear aloof, perched on the top of the ivory tower of medicine, apparently far out of reach except for the brief period of appointed time patients are lucky to get on the clinician's schedule. Social media helps bridge the gap between the needy patient and the hard-to-reach clinician. It is digital comfort food for many people who have nowhere to turn, apart from the bleached, sterile halls of managed care.

Social media can be used in clinics to enhance patient communication. There are several formats where recall can be effectively used or newsletters proliferated. Hy.ly is a platform where your staff can communicate with one another, patients can submit comments or complaints, and their input can be flagged for action so it is not left unattended. ePrescribing is gaining hold, and many clinicians already have a system in place that expedites the prescription from the electronic medical record directly to the pharmacy. You will have to ePrescribe to qualify for the federal Electronic Medical Record Reimbursement plan. As discussed throughout this book, social media is a fantastic and inexpensive way to create a public face and draw new patients, either via direct communication with a Facebook page or Twitter feed, or by blogging and creating content that helps your organic listing in search engines float higher and higher toward the first page. It can be used to enhance your professional reputation among your colleagues, but accessibility to you can enhance your reputation among your patient base as well.

Powerful Rules to Enhance Your Medical Practice with Social Media

Create a Patient Support Network

This is an excellent use of social media for a small medical business. Create a forum and discussion board for your patients to discuss relevant topics. Advertise it in the examination room and the waiting area. This will not only serve the sharing and disseminating of information by you, the author, it helps you to keep your ear to the ground, learning what your patients needs and concerns are, and sometimes giving you clues as to what you are doing right and what you are doing wrong as a small medical business. Encourage your patients to create topics; to start, you may want to incentivize patients to contribute by offering them a small perk, such as a $5 Starbucks card, just to get things going. Get your staff involved by contributing topics as well. Be sure to use the forum function in Google (type subject + forum, inputting your specialty in the subject phrase), and you will get a list of similar forums and discussion boards on the Internet. Peruse these sites and use similar questions and forum topics for your forum. Interact within those sites and market your forum by putting links to your content on other people's forums, as these are people with similar interests who may help you grow your forum faster. The links make you more relevant and float higher when potential new patients search for your specialty in your geographic region—it is what I call being searchial, not just social.

Pitch Yourself as an Expert; Thought Leadership in the Social Sphere

There are plenty of snake-oil salespeople on the Internet who, for a fee, offer to help you become a "thought leader." Speaking from personal experience, there is plenty of opportunity to establish yourself as a thought leader through social media for free. The formula is good content + linking + proliferation = thought leadership. The thought leaders on the Internet are those who create good content that people searching for particular keywords or key phrases in the search engine can find. The best search engines make the best content the easiest to find and reward the authors with better placement, cementing their reputation as thought leaders. There is plenty of information online, but as you enhance your image as an expert in a particular field by creating content, properly linking it, and proliferating it with tools like Twitter, Facebook, and social bookmarking sites, people will seek your opinion over others after reviewing other information out there, valuing it more than other content they found online. Put photos of interesting patient conditions on flickr.com, and share and discuss them with colleagues. Develop your own apps for PDAs, but most of all, work on content that you link to and post on your blog. This is the best place to share your expertise with the purpose of being recognized as a thought leader (see blogging, Chapter 9).

Methods of Increasing Patient Compliance Via Social Networking

Ensuring patient compliance is a major challenge for the healthcare professional. Efforts to increase patient compliance consume tens of millions of healthcare dollars each year, not to mention the time it takes during patient contact to get the compliance message across. For many years and even today, physicians depended on traditional marketing efforts to encourage patient compliance, such as providing brochures, pamphlets, fact sheets, and other written materials, all with a cost to the practitioner when ordering or printing these written materials.

Downloadable Materials

A step in the right direction, steering patients toward downloadable references or other web references that can be read on the computer screen or printed off, has been a strong trend over the past twelve years. Some forward-thinking practitioners transferred the compliance messages they used to get across in pamphlets onto their websites and refer patients to the URL.

Blogs

Blogs become helpful, as the information on blogs is quick and easy to change, add to, or update. Using downloadable materials can be a step in the right direction but usually involve expenses associated with web development and changes to the website. As fast as things change in the healthcare field, online pamphlets and informational packets may need regular updating, and the more updating, the greater the expense in terms of how much you have to pay your developers to make those changes. By putting patient information on your blog, you can use your content management system (editing feature common to all blogging software) to make changes yourself, or delegate to staffers to ensure your materials are the most up-to-date. Providing the blog's URL to clients and letting them know you update it monthly or quarterly reflects well on your practice; it lets your clients know you care enough to provide them with regularly updated information on the issues they are concerned about, and when they know you write a blog, it makes you appear current with trends in technology and puts a better face on your practice. There is a double purpose—the information you post for client education becomes another blog post that you can use as content to enhance your online presence, creating links that add marketing value to your business. Again, a recurring theme in this book and a very important message I want to get across is that the better, more relevant content that you post and get links to, the higher your blog floats in search engines when people in your geographic vicinity search for keywords or key phrases specific to the information you are supporting. Your efforts to inform and educate via your blog are simultaneously marketing your business online.

Videos and Web-Based Patient Education

The revolution of Internet video is here, and it's growing. Videos are pervasive on the Internet (see Chapter 12) and drive tremendous amount of search. Recently, YouTube became the number three search engine, after Google and Yahoo . Visual images are becoming preferred to text in many cases, and what better opportunity to educate patients than to provide video covering the medical subjects they care about. Creating video and incorporating video links on your blogs and websites helps your web efforts by increasing your relevance within search while simultaneously enhancing your professional reputation. In the section on video, I introduce you to some easy and intuitive tools to produce video for free, and provide links of videos I made for my practice by myself, some of them generating hundreds of visits in a short time. The social web is at the point where you need to use video or you are penalized; competitors who use video might rank higher than you in search-engine searches for the products and services you are offering. Just like with blogs, creating videos has three immediate benefits: (1) enhancing your ability to educate your clients while (2) increasing your marketing efforts by enhancing your reputation as your patients recognize and appreciate your video efforts, and (3) posting video improves your position in search engines.

Client Reminder and Recall Systems

I have heard of businesses using Facebook and Twitter as client recall systems. At the very least, I'm sure some of you reading this book use e-mail for the same purpose. The benefits of using social-media suites for recall is that people rarely change their "page" to another, so, for example, as your Facebook patient list grows, you are likely to retain most of the people within that list. On the contrary, people change e-mails frequently (such as when their job changes), and this can create a less-than-perfect recall list. Also, you can recall many patients at once using social-media suites, without one being aware of the other, and at the click of a button, whereas with e-mail you have to recall one at a time, and it can be a time-consuming process for your staff, not to mention all the bounced e-mails they will subsequently have to remove from your database. Facebook is better than Twitter for this, as you can set privacy settings to maintain patient confidentiality. HIPAA has yet to chime in on this, so it's kind of like the wild west right now, but I am certain that in the not-too-distant future there will be guidelines for patient communication via social media. Twitter is less private, and other than direct messaging your Twitter connections, it is viewable by most people and may have limited use. Tweets are on a stream, and if the person doesn't pay attention to it, it is not likely they will see it again, whereas in Facebook, the message remains in their inbox until they consciously delete it. There are, however, social media sites called aggregators, some of which allow you to program tweets, and the possibility exists for you to program in your recalls using tools like socialoomph (see Chapter 10) so they are automatically generated days or weeks ahead. I'm sure the tool to schedule a tweet a year ahead exists, but I haven't found it yet. When I do, I might experiment with it as a recall system. Think about how much postage and staff time we can save with the right social media tools. Schedgehog.com is a mobile app that can fill the cancelled or missed appointments in your schedule. Schedgehog is free for your clients; your staff loads cancelled or missed appointments into the database, and people seeking care at the last minute find the available appointments when they search by zip code—with one click, they are connected to your receptionist, and that lost revenue is converted into found revenue through last-minute filled appointments. Schedgehog membership is inexpensive; one patient a month who fills an empty slot can pay for your monthly membership fee three times over.

Use Social Networks to Share, Network, and Gain Support from Colleagues

Social networks are a great way to learn, and more professionals are using healthcare social networks as a source of information sharing. Fill your Twitter following list with influential clinicians and other healthcare professionals who post interesting and informative content. That way, every time you open your feed, some type of useful information is bound to jump out at you. Twitter is an excellent tool to use to stay on top of what is happening in the online medical community. Communicate with the experts; every time your name is seen messaging a world-renowned expert, people observing the conversations will look at the tweet and think, *Hey, who's that guy talking with Dr. So-and-so—he must be important! I'll connect with him!* Just by being connected to influential people, your influence can grow, and your follower list grows when your influence grows. Connect to influential healthcare experts on Facebook for the same reason. Professional journals are increasingly moving online, and coupled with professional continuing education, they are making staying current easier for professionals. Be sure to find the journals you like to read, subscribe to them, and check out the profiles of your colleagues on LinkedIn to see whom they are connected to and which journals they read. Medical portals, such as KevinMD.com, can introduce you to experts and conversations in the medical field that can increase your followers and followings.

Digg.com is an excellent place to explore to find content that other people like and have labeled as such, or to post original content that you hope to have recognized. On Digg.com, people vote, or "digg" their favorite content, and the more votes content gets, the higher it floats to the top of Digg, generating more interest and increasing your relevance in Digg searches. If you want the public to be aware of something you published, Digg it—it's an opportunity to see what interest is out there, and if there is a lot of interest, your content may become popular Internet-wide, even viral. (See Chapter 7 for more information about aggregators, social media tools that proliferate content.)

Many pharmaceutical companies are starting to create and maintain customer service portals that make it easier to communicate with the company and obtain things like point-of-service items, videos, patient education tools, among others. Some companies offer live video conferencing and continuing education online as well. e-Pharm alert is a good example—it enables communication with physicians via an automatic alert system that puts e-mails in their inbox when there is relevant information on particular pharmaceutical products, such as changes in physician prescribing information, recalls of certain products, etc.

Mobile applications are providing physicians with medical tools they can use when out of the office. Epocrates is a tool you can use on a PDA that updates regularly with the latest information on medical treatment for specific diseases and pharmaceutical information. ReadhMD CME is an iPhone app that allows users to listen to CME content and take a certification course for credit. Medscape has a free iPhone app that offers many functions, such as a drug database, drug interaction checker, and a health directory of physicians and pharmacies. ICD-9 is an iPhone app that allows you to query ICD-9 CM codes on an iPhone. MedCalc is a free mobile app across many mobile platforms that supports a large selection of formulas and scores for the physician. This year, I founded Schedgehog.com, a mobile app that lets doctors post cancelled appointments or recently opened appointment slots, so that busy people can go to the mobile app, type in a zip code and type of physician or other healthcare service they seek, and find an immediately available local appointment. The tool pays for itself if just one slot a month is filled, as missed and cancelled appointments are a major source of lost revenue for physicians. The tool is free for the general public, so is likely to be utilized liberally to the benefit of the participating medical practices. The user interface is friendly, so when a slot opens in your books, your receptionist goes directly to the user interface and enters the appointment which immediately shows up in the schedgehog database. The patient who needs a dental checkup in zip code 20850 simply enters the zip code and the specialty, and all open appointments in a nearby radius show up—one click and they're on the phone with the office reserving the slot. The mobile apps area is the largest-growing area of media and social media with application to the medical practitioner or allied health professional.

Here are some popular social networks for physicians:

- Ozmosis
- SocialMD
- Doctors, PAs and nurse practitioners networking
- Sermo

ADVICE

Always make people aware that what you post should not be construed as medical advice. Don't let this scare you; social media is here to stay, and doctors will be using it. Treat your social media efforts as curbside advice, but approach any posts on medical issues with caution. Use the advice in this book mostly to create content to help your patients, enhance your professional reputation, and benefit the growth of your practice by increasing your importance in search-engine rankings. Oh, and as you get more dependent on mobile phone apps for patient care, try not to lose your phone!

APPENDIX C:
SOCIAL MARKETING
FOR CLINICS AND HOSPITALS

Whether you are a clinic or hospital, employees, patients, and the community in general are talking about you online. Are you steering the conversation or sitting on the sidelines? Are you even aware of what is being said about you? For hospital administrators, there is a wealth of information to absorb and implement, you just have to keep your ear to the proverbial ground of the social Internet. In the book *Groundswell* by Li and Bernoff, they report that data shows hospitals implementing social media have a significant increase in web traffic. A spring 2009 Ad-ology Media Influence on Consumer Choice survey from Ad-ology Research reports, "Social media impacts nearly 40 percent of recent hospital or urgent-care center patients."

Digital Technologies and Medical Care

In this age of technology, patients are better educated about their medical condition and the methods utilized by healthcare providers to treat them. Digital, especially social media technologies, have enabled patients to become more participatory in their healthcare, enabling them to ask better questions. Many patients know where to find good information, enabling them to exercise more control over their healthcare and involve themselves more in the decision making. The digital revolution is moving people away from being passive receivers of healthcare into a more active role. If you aren't available online to answer their questions, someone else will be, and you might lose a patient, not to mention some ground in the "groundswell" of communication that is occurring online.

According to the Pew Research Center, in 2000, 25 percent of American adults looked online for health information. Now (2010), 61 percent of adults look online for health information. The Pew Research Study on 2,054 adult "e-patients" (patients using the Internet) ages eighteen and older, conducted by associate director Susannah Fox and published in 2008, provided some other interesting statistics:

- 93 percent of e-patients said it was important that the Internet made it possible to get the medical information they needed when it was most convenient for them.
- 92 percent said that the medical information they found was useful.
- 91 percent looked for information on a physical illness.
- 83 percent said it was important that they could get more health information online than they could from other sources.
- 81 percent said that they learned something new.
- 80 percent visited multiple medical sites. A few visited twenty sites or more.
- 72 percent searched for medical information just before or after a doctor's visit.

Digital technologies and social media forms are supporting and encouraging active involvement, and not just by patients but by all connected parties. Patients, caregivers, healthcare professionals, and pharmaceutical companies all are connecting as part of a larger ecosystem.

Ed Bennett, marketing professor and director of web strategy for the University of Maryland Medical Center in Baltimore, estimates that as of February, 2010, fewer than 14 percent of hospitals currently use social media. Even more shocking, he estimates that 50 percent of the hospitals in the US are still blocking access to social media and that there are some major roadblocks for mass adoption. He goes on to note that larger hospitals have a higher adoption rate of social media than smaller hospitals, and attributes the disparity to the lack of resources inherent in smaller healthcare institutions.

The following statistics were compiled by Ed Bennett in his Hospital Social Network Data & Charts. As of August 9, 2009:

- 253 hospitals have Twitter accounts
- The average numbers of followers on these accounts is 294, with a mean of 202
- First hospital with Twitter account was St. Jude Children's, October 23, 2007
- Most followers: Henry Ford Health, with 1489
- Most updates: Alegent Health, 1349 updates

- There are eighty-two Facebook accounts with an average membership of 821
- St. Jude's has the most fans: 33,252
- There are 121 hospital YouTube channels with a total 4,575 videos
- The first hospital YouTube channel was set up by Arkansas Children's Hospital on September 13, 2006
- Avera Health has 377 videos on its YouTube channel—more than anyone else
- M.D. Anderson Cancer Center has the most subscribers at 575

Public Health Education and Preventive Healthcare Using Social Media

One of the most popular healthcare topics is preventive healthcare. Having patients practicing preventive care benefits doctors and patients and generally makes our country healthier and stronger. Hospitals and clinics are places where much of the public gets educated on maintaining healthy lifestyles; unfortunately, by the time the patients receive the information, they are usually already admitted for conditions that may have been prevented had the patients been aware of how much their lifestyle contributed to their conditions. Providing preventive health information is a good place to get started on a social media campaign for a clinic or hospital. Your hospital should blog regularly about preventive care, such as smoking cessation and nutrition. Use the "events" function on Facebook to promote seminars about public health and preventative information for patients, or send out messages on Twitter. People are already getting bombarded with these messages online, so it is important your hospital or clinic gets in on the game to continue to be relevant in your community.

Connecting with patients on social media and providing educational tools keeps your organization in the public spotlight, reflects positively, and enhances your reputation as a healthcare thought leader. Think of the large market of potential customers you can draw if you can get the message across that your hospital is not only the place to go in the community when ill, but when seeking healthcare information or integrative medical services offered at your hospital; publicizing this could provide additional revenue. The more you can get in front of your community with good health information, the more likely they will think of you when the time comes to receive care, allied health services, or information on care. Social media provides many tools for regularly updating information on preventive healthcare. Having your professional staff create original content in their particular area of expertise and linking that content to other important healthcare sources on similar information, such as the NIH, FDA, WebMD, and many other prominent healthcare sites, will boost your hospital's search-engine ranking, helping your business while helping patients at the same time—a win-win using social media.

Opening Patient Communication Channels through Blogging

Social media in healthcare is about clinics and hospitals extending care and commitment to community healthcare beyond their walls, engaging the public with the goals of listening, learning, educating, and marketing. Using social media this way can be used to create a significant patient channel. Social media provides many different ways for clinics and hospitals to communicate with the public.

One of the most powerful methods a hospital or large clinic can utilize is blogging. Successful blogging as a healthcare organization involves making available the information your community seeks, and that the information is timely, updated, and readily available in their time of need. Blogging is popular because of the ease with which one can post, edit, and disseminate information. It keeps your information fresh and in the public eye. A popular blog can do wonders for a clinic or hospital and is viewed by the public as "patient friendly," as it is easy to reply to or post comments on blog posts. It is important, however, that the hospital maintain the blog and communicate back with the public—use the blog as a PR tool; medical information is one thing, but updates on projects, charities, and events the hospital or clinic is hosting can be marketed alongside medical information. Writing a blog with depth, one that covers numerous topics, both serious and lighthearted, is the key to a blog's success. Remember to write *to* your audience, not *at* them. Hospitals already have enough of a challenge fighting the perception of being higher and mightier than their communities and are always facing marketing and customer-service challenges due to this reputation. Blogs are an excellent tool to make your organization appear friendlier and more patient-centric and can be implemented virtually cost free. Be sure to read the chapter on blogging.

YouTube and Hospitals

YouTube is by far the most frequently implemented social media tool for hospital marketers. This is likely due to the fact that video has been an important marketing tool for hospitals for more than twenty years now. The advent of easy-to-use digital cameras with instant uploads to the Internet enables an efficient and inexpensive means for getting a message across to a large patient population. Video also lessens the amount of time staff spend educating patients, and the production and editing of video enables more thorough information to be provided to patients than their doctor often has time to provide face-to-face. Video can be used to mass introduce new medical technologies that may provide a PR and profit benefit to the hospital that invests in them. Links to these videos are easily disseminated through social media and e-mail for ease of access by patients. Videos proliferated via social media that introduce the expertise of hospital staff and patient-friendly policies can help a hospital solidify itself as a thought and care leader in the neighborhood or city in which they provide care.

LinkedIn and Hospitals

LinkedIn, discussed earlier in the book, is a tool professionals use to connect and share. The medical profession has much to benefit from being connected to a vast network of similar-minded individuals. In the past, most sharing of medical information occurred through print journals and conferences. Now, links can be shared with colleagues, notices about webinars (online seminars) and conferences shared, and finding the colleagues one should be sharing with is made easy with social media networking tools like LinkedIn. As a doctor's connections on LinkedIn grows, connecting with other doctors becomes easier. Sharing of information also becomes easier, which ultimately leads to improved patient care. Imagine a patient with a rare condition, and how challenging it might be to locate the specialist with the most experience to treat that patient. Now imagine how difficult it might be to actually connect with that specialist. On LinkedIn, you can see which connections lie between you and the specialist and use common connections to get in touch faster and easier to get the information you seek. (See LinkedIn, Chapter 10.)

Patient Care Websites: Carepages.com and Caringbridge.org

Caringbridge and Carepages are nonprofit organizations that act as blogs for people facing serious medical conditions, hospitalizations, or are recovering from illnesses, accidents, or medical procedures. The patient posts updates to inform family members and friends of their struggles, progress, and recent medical information, eliminating the need to be in contact with everyone concerned via e-mail and phone calls. Visitors given a code receive the updates and can post their own messages of care and concern that the patient reads and goes to for support. Caregivers can participate and add to the conversation with medical information if they choose. Followers are notified via e-mail with updates. Caringbridge is not searchable by major search engines, and the names of the patients can be kept private. As of September 2009, more than 160,000 people have created a CaringBridge website, totaling more than one billion visits by friends and family. Both sites have additional links, and content on medical treatments, advances, and news patients can use as a reference.

There are many other helpful websites and social media sites geared toward patient care that I'm not going to dedicate space to here, but many can easily be found with a Google search. As a physician or allied health professional, you may opt out of participating in social media, but in order to stay relevant and come across as someone who is in touch with the needs of your patients, it is imperative you stay abreast of the information sources available to your patients online.

Clinic and Hospital Social Media Strategy, Planning, and Policy

A social-media strategy document for a healthcare organization should be short and succinct; there's no reason to make it longer than a page or two. The first subject that should be addressed is privacy concerns, and it should be made clear to all employees that any and all social interaction is bound by the rules of HIPAA. Discussions should be limited to work-related subjects, and policy should state that off-topic conversation will be deleted.

Final copy of content, such as in blogs, should be required to be approved by the public relations or marketing departments prior to posting. It should be made clear to everyone involved in the social media campaign that there is zero tolerance in regard to posting of false or inaccurate information. Linking should be scrutinized, with only high-level authority links created and posted, eliminating links that may be considered spam.

An online-interaction culture should be encouraged. The online conversation should take place with the outside world, and tools should be implemented to enable an internal social network so people continue the culture throughout the work day. This will add to the effectiveness of social media implementation. A great tool for interoffice communication is Hy.ly, which also has a module to monitor conversations on the Internet about your clinic and receive, log in, and communicate patient comments, suggestions, and/or complaints. A simpler version of Hy.ly is Yammer, which is basically an interoffice version of Twitter; if you are the business owner, the conversation stream between your staff can appear on your smartphone or e-mail stream so you can oversee goings-on at your office when you are away.

Social media is helpful in hiring. LinkedIn is a particularly good way to network to find new professionals who may be job seeking, and Twitter is inundated with medical placement services sending out tweets of available positions to be filled. There are many opportunities to find experts in various clinical areas, interesting speakers, advisors, or consultants. Relationships should be built with employees and clinicians at other hospitals so you can learn which tools other organizations use in social media and how they use the tools administratively as well as in their patient and networking communications.

Forums and discussion boards are the oldest social media tools. They are incredibly useful social media tools for hospitals. Not only can they provide medical education, interaction with other colleagues, and general information for physicians, they can be used interoffice for marketing, communications, and business development. They are fantastic archival sources, so the information posted can be easily catalogued and referenced later. People can view the posts to see which were the most popular, so they can go directly to the probable better sources of information. They can participate across medical facilities to expand their reach, thus the depth and breadth of the information they provide. They are great places to exchange ideas, engage in healthy debate, and present new ideas. Encourage the online communities in forums and discussion boards to participate by listening and interacting. On forums, discussion boards, and social media in general, it's not "set it and forget it"—you must post, read posts, and respond to posts; it's a good habit to get into to make social media pay off for you.

Excellent examples of communities exist in my industry. Sightnation.com is a social media portal which includes blogs, forums, and discussion boards for the optical industry. I highly recommend visiting the site to get ideas about how you can develop a social media portal in your industry.

Enhancing Your Brand within Your Community

Social media is the perfect tool to build goodwill within a community. Efforts within social media to do good draw interest, benefiting the recipients as well as the organization offering the goodwill. Stories about instances of special care given by a department or an employee, of rare medical cases treated, or patients who couldn't afford care having it donated reflect positively on your organization. The PR aspect of social media is exceptionally powerful for spreading word of your good deeds, quickly, inexpensively, and effectively and should be utilized; just be careful not to overdo it. Remember also, hospitals historically are businesses the community prefers not to interact with, yet the more interaction the hospital has with the community, and the more patient-centric it appears, the better for business, so part of any hospital or clinic branding strategy should incorporate the positive PR social media can provide. Hospitals have an opportunity through social media to re-brand themselves away from the image of a place you go to wait, receive poor to average care, or go to only when you are sick, into an organization that provides social good, community activities, allied health services, and is perhaps a sounding board for personal and community ills. The hospital of the twenty-first century should strive to be patient-centric, provide education and community outreach, and re-brand themselves as such. They can start by interacting with their public through social media channels.

Public Image Enhancement and Damage Control

On today's Internet, everyone and anyone can come under attack, and the party being attacked is usually the last to know it. By the time you are aware of something damaging being said about you in social media, it may have already circulated around your community. The only way to protect yourself from this type of attack is to try to consistently offer an exceptionally high level of customer service and be involved in the conversation and monitor what is being said. Defending yourself, and taking the high road when doing so go a long way toward reputation damage control.

It is important to know the most popular review sites and scour them periodically for mention of your business. Most review sites provide a space next to the negative comment to rebut. Be sure to do so professionally. If the negative comment is posted on a review site, be sure to find customers/patients who are happy with your services and ask them to post on the same site, thereby diluting the negative comments to future viewers. If the party responsible for posting the negative comment is reachable, reach out to them and see if they are open to discussing the problem. Put your customer service hat on and see if there is anything you can do to make it good by them. Many review sites will allow the poster to remove their negative review. Be sure to have a system in place to promptly deal with the comment—the longer the post has been up, the less likely the poster will be to remove it. Demandforce.com is a novel platform for medical professionals through which you can monitor online reviews and act on them. The suite of tools they offer has an impressive amount of depth, and I use it regularly to monitor reviews of my practice. (If you mention *Searchial Marketing*, Demandforce will give you two free months when you sign up. Contact them for a free online demo.

Interacting with the public is a tremendous way to enhance the image of the clinic or hospital you represent. Most people only connect with hospitals when they are sick, and hospitals have a reputation for being cold places with unfriendly receptionists and long waits. Social media offers a way for hospital employees to get beyond the perceived cold walls and get out into the public to demonstrate that most hospitals and their employees are there to help, and that helping can extend to the neighborhood. Finding creative ways of getting people into the hospital for classes or other events also enhances a hospital's reputation, and social media is an excellent and free PR tool to help disseminate information on public relation events, enhancing the hospital's reputation within the community as a friendly and helpful place to go to even when not in need of emergent medical care.

Hospitals can respond to negative media reports and coordinate damage control with social media. Press releases can be issued in response to these reports, but the "sharing" function available on most websites and blogs is an effective way of quickly countering negative reports and disseminating information on the topic to large groups of people.

Other uses

Hospital events can be organized through Facebook's events function. Almost all tools, especially Twitter, can be utilized when staff or patients want to coordinate gift giving or jointly send gift baskets or flowers to patients. For patients with loved ones who have been admitted, many of the tools are useful for providing updates on the patient's status (within the bounds of HIPAA, of course), instead of them having to call everyone and waste hours on the phone. This duty can be delegated to hospital staff, such as nurses, during their rounds.

Here are examples of clinicians, clinics, and hospitals effectively using social media that you can learn from.

- RunningAHospital.com—Beth Israel Deaconess & Harvard Medical Center
- Stjudefriends.org/socialmedia/—St. Jude's Children Social Media Tool Kit
- Sharing.mayoclinci.org—Mayo Clinic's blog
- @MDAndersonNews—Twitter account
- Youtube.com/user/UMMCVideos—University of Maryland Medical Center You Tube Channel
- @SeanKhozin—An internist who uses e-mail, IM, and video chat for coordinating care

Make social media available on a hospital-wide basis. Use internal social media tools for staff to communicate, such as Hy.ly (see Chapter 14) and Yammer.com; encourage professional communication via sites like LinkedIn; and communicate with the city or neighborhood you serve via your blog. Encourage the community to connect *and* communicate with you on your Facebook fan page or through Twitter. Most importantly, listen to and address these communications—after all, *social* marketing involves being social; without the organization, you are not maximizing your use of this all-important tool. Lastly, use the strategies for content creation and proliferation throughout this book to get your message out, and participate in the conversation. Remember, as a hospital or large clinic, your customers are usually your neighbors and friends, so don't forget about the social aspect involved in social marketing and your efforts are likely to bear fruit—and fruit baskets!

APPENDIX D: SEARCHIAL MEDIA FOR BIOTECH AND PHARMACEUTICAL CONCERNS

Fortunately so far for pharmaceutical, biotech and medical device companies, the FDA hasn't directly addressed the issue of using social media to market drugs, but it doesn't categorically prohibit pharmaceutical companies from engaging in social media. The position of the FDA is likely, in my opinion, to be "it doesn't matter how you market, it's how truthful and ethical your marketing techniques are."

I recommend your company start by identifying the discussion board and message boards that exist in the particular market the company is competing within. Create a team to monitor the boards for mentions of company products, disease states that comprise the company's markets, customer opinion leaders (COLs)—the people who are the most active on the relevant forums and could possibly be recruited to effectively spread your message—and comments on competitors' products. Staff should consist of highly trained experts on the particular medical topic that can engage and interact within the forum and monitor the message board posts for valuable content. Assign a marketing and advertising expert to find the best forums to post advertisements, discounts, and coupons. Media placements and sponsorships within relevant communities is still an important strategy; while the typical small-business user of social media may want to decrease his or her advertising costs, the pharmaceutical concerns with deep ad and marketing dollars can still reap a significant benefit to these firms from traditional advertising. Placement should be a goal; deliver the right message to the right group.

Monitoring Company Products

Social media is a world where reputation rules and trust is king. Honesty is recognized and rewarded as well; attempts to skirt important issues or to control the conversation can be the cause for significant backlash. Thanks to this new world, branding is being reshaped within the digital realm. In this new social world, it is the swell of people discussing your brand that is shaping it, not traditional marketing and advertising. A company is no longer in control of its brand; the brand now controls the company. If the product holds its own in clinical trials and successfully completes the new drug application (NDA) process, the world will spread via physician forums and discussion boards quickly, and if there are problems with the product, it can die just as fast a death. The company that has problems in a clinical trial needs to be proactive within the discussion boards and forums and provide an honest explanation behind the failures that led to the problems. The company that attempts to hide the real reasons behind their failure will be subjected to scrutiny surrounding their entire product line, while the company that quickly and directly addresses the real issues will gain respect within the community, and people in the social community will likely be willing to reconsider the product or brand in the future. Because of this, it is crucial to actively participate in the conversation surrounding your brand, but more important, keep your ear to the ground and listen to the "groundswell" surrounding your product before involving yourself in the conversation. With free and accessible sources for information, such as discussion boards, message boards, and forums, the information you need to understand how your brand is perceived is out there.

Traditionally, pharmaceutical companies ("big pharma") have kept a short leash on the dissemination of information surrounding its brands and products. With the changing face of media and the advent of the blogosphere, it became difficult to keep important information regarding brands and products within the company. People seek information, and when they can't find it, will question why. This creates the need for pharmaceutical companies to increase transparency in order to maintain public trust. Online, the effort to withhold information is transparent and can lead to a significant backlash from people participating on the social web. The negative publicity the company experiences from this backlash is much harder to overcome than the act of providing the information would have been in the first place.

Getting the "Inside View" on Specific Markets

Social media sites can give pharmaceutical companies an inside view as to what physicians are thinking and saying about particular diseases and treatments. Through SM, it's finally possible to be the proverbial fly on the wall. Using the appropriate staff, scour the message boards, discussion groups, and forums to find where people are talking about the disease states relevant to your market. Listen and engage; promotion is okay, but only promote 20 percent of the time. No one participates in social media to have to dodge spammers or "walking billboards." Seek out people tweeting about the relevant disease states, and monitor the Facebook groups. Remember, each disease state has many discussion groups and forums out there. Finding them is easy; on Google you can type [disease state] + forum in the search bar, and every indexed forum relevant to the disease will come up.

Searching forums and discussion groups to monitor what is being said about your products, and participating in the discussion to add value and correct any misperceptions, is a cost-effective way to enhance your brand. The physician forum Sermo has an interesting product, the Dashboard, where clients can set up keywords that trigger hits whenever one of the 110,000 physicians mentions the particular product, disease state, or other term your company wants to monitor.

Information from feedback forums is different from the information gleaned from focus groups. Doctors and other allied healthcare professionals are much more likely to share their honest opinions online than they would in focus groups where they are being paid and feel an obligation to steer their opinions toward the favorable side of things. The right questions can provide information as valuable or even more valuable than focus groups, and at a fraction of the cost. Sales and marketing in the pharmaceutical industry is very expensive—social media can dramatically decrease sales and marketing expenses; the tools are out there and the conversations are ongoing; the onus is on the company to find out what is being said.

Competitive intelligence firms like Nielsen BuzzMetrics and Cymfony offer tools that collect detailed info about what consumers and medical professionals think about medications and related issues. This information is gleaned via user-generated media like forums and discussion groups.

Customer Opinion Leaders

Customer opinion leaders are the experts in a given field who are actively engaged in their particular online conversation. By liberally participating in forums, discussion groups, and social media sites, where the conversation surrounds their particular areas of expertise, they become valued thought leaders who draw people's eyes and ears to the content they produce, the questions they answer, and the opinions they produce. They can influence entire communities with just a line of text. It doesn't take much for a pharmaceutical company to recognize the value these thought leaders bring to the table. It is helpful to identify the COLs in the industry you compete in and subtly engage them in conversation, build a relationship with them, and gain their trust. Be honest and upfront early on as to who you are and what products you represent; they will appreciate knowing whom they are dealing with. Find out if there are ways you can help them disseminate whatever message it is they may hope to disseminate, if there are particular causes they are involved in that you can help with, or try to identify any way at all to "get your foot in the door" with the COLs. Social media is about helping others, and others expect to help you back; find the right COL with the opportunity to add value to his or her conversation, and you may gain an invaluable ally with a loud voice within a particularly important community for your business.

Comments on Competitors' Products

Competitive intelligence is another game that is played within social media. You can glean information on your products and services from forums and discussion groups, and you can glean information in equal amounts regarding competitors' products and services. To add depth to your efforts within these tools, it is important to record and analyze what is being said about your competitors within the same forums you participate in. Finding information on reputation and products happens in the forums. Chapter 13 describes various online tools you can use to "spy" on your competitors' Internet marketing efforts.

Some well-known online healthcare communities you can start with:

- Medscape
- WebMD
- iMedExchange
- PatientsLikeMe

Community Engagement to Increase Word-of-Mouth Referral

CooperVision, a major contact lens manufacturer, launched an impressive effort to market contact lens products toward a younger, more Internet-savvy consumer in their portal mycontactsports.com. The portal has contests and promotions and hires famous athletes to promote the use of contact lenses for sports. CooperVision does a great job of providing useful, entertaining content to a market segment they hope to gain, and they make it fun, social, and participatory. For doctors who use CooperVision products, they offer expert social media consultants who help the doctors learn about social marketing and even help build Facebook pages. CooperVision isn't just out there talking and posting content, incorporating keywords and key phrases clients use when searching for contact lens information, they are engaging the audience and creating a buzz surrounding their products. Charlene Li and Josh Bernoff, in their well-known book *Groundswell,* discuss how social media can "energize the base." They explain that "energized customers become viral marketers, spreading brand benefits to contacts without any cost to the company. Word-of-mouth is a powerful amplifier of brand marketing, achieving results no media campaign can achieve." This is exactly what CooperVision is doing with mycontactsports.com. Medical device and pharmaceutical companies need to consider this type of social interaction to engage and market within social media

Social Commerce

In social *commerce*, conversations with and between customers are enabled directly on business websites. In social commerce, customer conversations directly affect conversion—and drive revenue. Having real customer ratings, reviews, answers, and stories directly on the site where you are promoting your products and/or services is an extremely powerful marketing tool, much more so than the best marketing copy.

Companies like Bazaarvoice can implement ratings and review systems on websites, making it easy for customers to give feedback, ask questions, and share stories. Having these systems in place helps create a client "culture," creating a buzz about your brand while enabling companies to respond faster to customer needs.

Wikis and Pharmaceutical Companies

From Wikipedia (http://en.wikipedia.org/wiki/Wiki): A wiki (/wɪki/*WIK-ee*) is a website that allows the easy creation and editing of any number of interlinked web pages via a web browser. Wikis support multiple contributors with a shared responsibility for creating and maintaining content, typically focused around text and pictures. Wikis are typically powered by wiki software and are often used to create collaborative websites, to power community websites, for personal note taking, in corporate intranets, and in knowledge management systems. *Wikis* are simple web pages that groups, friends, and families can edit together.

Pharmaceutical companies should consider posting a wiki for each brand they market, as well as one for discussing their corporate history and presence. The administrator, who is the wiki creator, can add appropriate content, or delete content he or she feels is inappropriate, should someone post something untrue or libelous. The wiki allows people to form a community and creates value for the product as it builds, while maintaining modest control about what is said about the product. People appreciate the efforts of a company that creates information surrounding its product, particularly open-source information, such as wikis. While wikis are not known for providing absolute truth or perfectly accurate information, they are, nonetheless, consulted as a trusted source of information.

The purpose of the wiki when it comes to building a brand is to involve visitors in an ongoing process of creation and collaboration. It is a way to engage customers and involve them in the creative and branding process and an indirect way to glean valuable insight into how your product is perceived. Because wikis are so effective at enhancing collaboration, they are catching on with lots of companies.

Historically, the medical community has been held to higher standards when it came to the dissemination of healthcare information. The higher standards within our community have led to the creation of expert-moderated wikis. Some wikis allow one to link to specific versions of articles, which has been useful to the scientific community, in that expert peer reviewers could analyze articles, improve them, and provide links to the trusted version of that article. Expert moderated wikis are good places to identify COLs, and helping them with their wiki is a great way to build a relationship with them. Here is a list of medical wikis as examples: http://davidrothman.net/list-of-medical-wikis/

Facebook and YouTube

Facebook offers groups and pages to market your brand or product. Pages are more advantageous for business, and as such you should be sure to create a page for each of your products, and a corporate page as well. The value in creating a Facebook page comes from disseminating information to a network of people who subscribe to your page and engaging them in conversation. It is helpful to post links to other relevant sources of information, including video. Keep the conversation alive by providing relevant, interesting, and up-to-date content. Use the link function to drive traffic to your blog. Facebook makes it easy to create applications that subscribers can enjoy interacting with through their "Facebook Mark-Up Language" (FBML).

On our Facebook page, we have created a quiz application that we named "Determine Your Risk for Developing Macular Degeneration." The quiz spits out a risk level—low, moderate, or high—and upon viewing the results, I offer links that encourage people to connect to our Facebook group where we discuss eye- and vision-care issues. I use the quiz to find the target market, and when I have engaged them, I offer the opportunity for them to connect with our eye-care community, help them by providing good content, and grow my practice Facebook presence in the meantime.

Facebook also offers the opportunity to draw eyes and ears by performing social good. I discussed earlier in the book my Facebook group "Shady Grove Eye Care—Just Joining Provides Free Eye Exams for People in Need." For every one hundred people who join, I provide free eye care to someone in need. I hired a professional videographer to "tell" the story of the person receiving the eye care and post the video on YouTube, linking to it. The video is compelling and encourages more people to sign up with the Facebook group. Check out the video of our first person to benefit from the free care at this YouTube link: http://www.youtube.com/watch?v=ZC47TGfUfUE.

Just think about the opportunities for a company promoting a diabetes or arthritis product to promote a video featuring a patient who suffers from one of these conditions, walking viewers through the daily struggles of people with chronic conditions, and then showing their quality of life improve after treatment with their product. Pharmaceutical companies should take advantage of the opportunities to promote their medical products by inventing creative charitable efforts and promoting videos empathetic in nature.

Remember, however, that as useful as Facebook can be for interacting within your population of connections, it is not followed by search engines and has no value when it comes to adding points for your business when people search for your offerings in Google or other search engines.

Twitter Feeds and Bloggers' "Customer Opinion Leaders" (COLs)

COLs are out there on Twitter, and they're easy to find. Find them, watch them, listen to them, and check out their links and videos because they provide a wealth of information. Engage them when you have something to say, as their social media currency is "greener" for your organization than that of your average follower. Frequently access the external Twitter search engine http://search.Twitter.com. Enter your brand and corporate names to see who is steering the conversation regarding your brand, and find out which direction it is headed in. Before you tweet your message, develop it and make it discreet; nothing is more annoying on Twitter than getting hit up by spammy-looking advertising tweets. Use hyperlinks within your tweets to draw people to your other social content, specifically your blog. Twitter is like blog medicine, and tweets are the pills that drive people to the blog, making it healthier.

Be sure to subscribe to the popular blogs about your products and services, and watch them with a reader feed like Google Reader. You can find them easily by using the blog search function in Google—just type in the keyword you are searching, add the + sign and the word "blog," like this: Avandia + blog. This will pull up all the indexed blogs discussing Avandia. Comment on blogger posts and ask them if they would like a link from you. There is no better currency for a blogger than "link love"—the more external links the blog gets from reputable sources, the better the blog performs, a win-win for your products and the blogger, and you might just win the respect and loyalty of a small-time blogger when he gets contacted by a big company asking to link to him. That same small-time blogger may turn out to be a significant COL/influencer in as short a time as a year or two, so use all kind of blogs to plant your product "seeds." This is the kind of grassroots social interaction big pharma should be undertaking to help boost their reputation and that of their products through my searchial media strategy.

Clinical Trial Sponsors

One of the significant challenges you face as a pharmaceutical corporation is finding clinical trial participants, and finding ones who are likely to complete the trial. What better place than social media forums, groups, or pages to find large, targeted groups of people where certain characteristics desirable to your study can be determined by visibly listed interests, ages, and group affiliations posted on the site? These groups enable certain levels of pre-qualifying before they are even contacted. Save some money there? You will likely also save some more money by targeting the people within a certain region, as distance from a research site is by far the number one factor in the patient's choice whether to participate in a clinical trial, minimizing dropout and risk for the study. Social media used in this fashion can lower the overall cost of the study per randomized patient, and can help you to keep in touch with trial participants as well.

It's not likely that recruitment for studies, whether online or by more traditional methods, will matter to Institutional Review Boards (IRBs), as long as your study uses sponsor- and IRB-approved content to recruit. In order to find the right candidates to participate in your clinical trial via social media, it's important to define and know what the structure of the target patient group should be, and then structure a strategy that will incentivize them to respond to your request. Of course, your messages will have to be approved for inclusion on a social media site by submitting to an IRB, but once approved, you have your message and can get ready to systemically introduce it to various social media sites.

Other Uses

Pillbox is a government initiative from the National Library of Medicine (NLM) at the National Institutes of Health (NIH) and the Food and Drug Administration (FDA) with the goal of transforming the way pharmaceuticals are labeled in the future. Every year, more than a million calls are placed to poison control centers with requests to identify pills. Handling of these calls is expensive. Pillbox uses open development of social applications in an attempt to find a better way to label medications, increasing the ease of identification by the average citizen. Open development (also known as open source) occurs when organizations expose their project on the internet and seek help from the general public in creating code that enhances new or existing applications on the Internet, on smartphones, etc. Using open source coding can result in out-of-the-box improvements on existing ideas, as well as tremendous cost savings for developing new ideas. Pillbox is an application that was built with help from the community to help categorize and label pills using high-resolution images. These images, when standardized and approved by the manufacturers of the pills, could be added to drug labels by the FDA and eventually may be uploaded in the Pillbox platform, which in turn can use digital recognition via smartphone cameras and other means to help quickly and efficiently identify pills.

The trend toward smartphone-based mobile applications is the fastest growing area of the Internet as of this writing. The possible applications are endless, and the field is relatively new. The potential for a company to develop an application that can drive product and profit or provide benefits to doctors or patients is enormous. Think of the many ways mobile applications can be used to create efficiencies and profits in your biotech, medical device, or pharmaceutical company. I suggest creating a team to implement and explore ideas for mobile applications that can give you a leg up in your industry. Now's the time.

Medical directories you'll want to be listed in:

- MedMark.org
- Healthdirectorymoz.com/Web-Directories
- Google.com/Top/Health
- Doctor.webmd.com
- Medicaldirectoryonline.com
- Callmedic.com
- Dir.yahoo.com>health

BIBLIOGRAPHY

Chapter 1
Li, Charlene and Josh Bernoff, *Groundswell: Winning in a World Transformed by Social Technologies*, 7. Boston, Massachusetts: Harvard Business Publishing, 2008.

Chapter 5
Li, Charlene and Josh Bernoff, *Groundswell: Winning in a World Transformed by Social Technologies*, 24. Boston, Massachusetts: Harvard Business Publishing, 2008.

Chapter 6
Gillin, Paul, "Social Media Marketing Still Lacks Strong Metrics." April 1, 2008. *Journal of New Communications Research*, Vol. II. Retrieved on November 2, 2008; http://en.community.dell.com/dell-blogs/b/direct2dell/archive/2006/07/13/flaming-notebook.aspx

Chapter 7
Comm, Joel, *Twitter Power 2.0: How to Dominate Your Market One Tweet at a Time,* 165. Hoboken, New Jersey: Wiley, 2010 (revised edition).

Chapter 8
Gillin, Paul. "Social Media Marketing Still Lacks Strong Metrics." April 1, 2008. *Journal of New Communications Research*, Vol. II. Retrieved on November 2, 2008. http://www.marketingcharts.com/direct/social-media-marketing-still-lacks-strong-metrics-4039/

Chapter 10
Li, Charlene and Josh Bernoff, *Groundswell: Winning in a World Transformed by Social Technologies*, (p.). Boston, Massachusetts: Harvard Business Publishing, 2008.
Brogan, Chris, *Social Media 101: Tactics and Tips to Develop Your Business Online*, 55. Hoboken, New Jersey: Wiley, 2010.

Appendix 3

Li, Charlene and Josh Bernoff, *Groundswell: Winning in a World Transformed by Social Technologies*, p. 24. Boston, Massachusetts: Harvard Business Publishing, 2008.

Ibid., 22.

Fox, Susannah, "The Engaged E-Patient Population: People Turn to the Internet for Health Information when the Stakes are High and the Connection Fast," Pew Internet & American Life Project, August 26, 2008 htttp://uploadi.www.ris.org/editor/1232628818PIPHealth Aug08.pdf

http://www.markeingvox.com/1-in-4-hospital-urgent-care-patients-influenced-by-social-media-043892/

http://www.pewinternet.org/Reprots/2007/Epatients-With-a-Disability-or-Chronic-Disease.aspx

http://wascom/impressions/2010/03/11/hospitals-social-media-better-care

http://staffingrobot.com/staffingrobot/2010/02/hospital-social-media-adoption-data.html

Appendix 4

Li, Charlene and Josh Bernoff, *Groundswell: Winning in a World Transformed by Social Technologies*, 130. Boston, Massachusetts: Harvard Business Publishing, 2008.